A MANUAL

OF

ELEMENTARY

GEOMETRICAL DRAWING,

INVOLVING THREE DIMENSIONS.

DESIGNED FOR USE IN HIGH SCHOOLS, ACADEMIES, ENGINEERING SCHOOLS, ETC.;
AND FOR THE SELF-INSTRUCTION OF INVENTORS, ARTIZANS, ETC.

In Five Divisions.

DIV. I. ELEMENTARY PROJECTIONS.
DIV. II. DETAILS OF CONSTRUCTIONS IN MASONRY, WOOD, AND METAL.
DIV. III. RUDIMENTARY EXERCISES IN SHADES AND SHADOWS.
DIV. IV. ISOMETRICAL DRAWING.
DIV. V. ELEMENTARY STRUCTURAL DRAWING.

BY S. EDWARD WARREN, C. E.

PROFESSOR OF DESCRIPTIVE GEOMETRY AND GEOMETRICAL DRAWING IN THE RENSSELAER
POLYTECHNIC INSTITUTE, TROY, N.Y.; AND AUTHOR OF A TREATISE ON THE
ORTHOGRAPHIC PROJECTIONS OF DESCRIPTIVE GEOMETRY.

British Library Cataloguing-in-Publication Data
A catalogue record for this book is available from the
British Library

Technical Drawing and Drafting

Technical drawing, also known as 'drafting' or 'draughting', is the act and discipline of composing plans that visually communicate how something functions or is to be constructed.

It is essential for communicating ideas in industry, architecture and engineering. The need for precise communication in the preparation of a functional document distinguishes technical drawing from the expressive drawing of the visual arts. Whereas artistic drawings are subjectively interpreted, with multiply determined meanings, technical drawings generally have only one intended meaning. To make the drawings easier to understand, practitioners use familiar symbols, perspectives, units of measurement, notation systems, visual styles, and page layout. Together, such conventions constitute a visual language, and help to ensure that the drawing is unambiguous and relatively easy to understand.

There are many methods of constructing a technical drawing, and most simple among them is a sketch. A sketch is a quickly executed, freehand drawing that is not intended as a finished work. In general, sketching is a quick way to record an idea for later use, and architects sketches in particular (in a very similar manner to fine artists) serve as a way to try out different ideas and establish a composition before undertaking more finished work. Architects drawings can also be used to convince clients of the merits of a design, to enable a building constructer to use them, and as a record of completed work. In a similar manner to engineering (and all other technical drawings), there is a set of conventions (i.e particular views, measurements, scales, and cross-referencing) that are utilised.

As opposed to free-sketching, technical drawings usually utilise various manuals and instruments. The basic drafting procedure is to place a piece of paper (or other material) on a smooth surface with right-angle corners and straight sides – typically a drawing board. A sliding straightedge known as a 'T-square' is then placed on one of the sides, allowing it to be slid across the side of the table, and over the surface of the paper. Parallel lines can be drawn simply by moving the T-square and running a pencil along the edge, as well as holding devices such as set squares or triangles. Other tools can be used to draw curves and circles, and primary among these are the compasses, used for drawing simple arcs and circles. Drafting templates are also utilised in cases where the drafter has to create recurring objects in a drawing – a massive time-saving development.

This basic drafting system requires an accurate table and constant attention to the positioning of the tools. A common error is to allow the triangles to push the top of the T-square down slightly, thereby throwing off all the angles. Even tasks as simple as drawing two angled lines meeting at a point require a number of moves of the T-square and triangles, and in general drafting this can be a time consuming process. In addition to the mastery of the mechanics of drawing lines, arcs, circles (and text) onto a piece of paper – the drafting effort requires a thorough understanding of geometry, trigonometry and spatial comprehension. In all cases, it demands precision and accuracy, and attention to detail.

Conventionally, drawings were made in ink on paper or a similar material, and any copies required had to be laboriously made by hand. The twentieth century saw a shift to drawing on tracing paper, so that mechanical copies could be run off efficiently. This was a substantial development in the drafting process – only eclipsed in the twenty-first century with 'computer-aided-drawing' systems (CAD). Although classical draftsmen and women are still in high demand, the mechanics of the drafting task have largely been automated and accelerated through the use of such systems. The development of the computer had a major impact on the methods used to design and create technical drawings, making manual drawing almost obsolete, and opening up new possibilities of form using organic shapes and complex geometry.

Today, there are two types of computer-aided design systems used for the production of technical drawings; two dimensions ('2D') and three dimensions ('3D'). 2D CAD systems such as AutoCAD or MicroStation have largely replaced the paper drawing discipline. Lines, circles, arcs and curves are all created within the software. It is down to the technical drawing skill of the user to produce the drawing – though this method does allow for the making of numerous revisions, and modifications of original designs. 3D CAD systems such as Autodesk Inventor or SolidWorks first produce the geometry of the part, and the technical drawing comes from user defined views of the part. This means there is little scope for error once the parameters have been set.

Buildings, Aircraft, ships and cars are now all modelled, assembled and checked in 3D before technical drawings are released for manufacture.

Technical drawing is a skill that is essential for so many industries and endeavours, allowing complex ideas and designs to become reality. It is hoped that the current reader enjoys this book on the subject.

CONTENTS.

DIVISION THIRD.

RUDIMENTS OF SHADES AND SHADOWS.

DIVISION FOURTH.

ISOMETRICAL DRAWING.

DIVISION FIFTH.

ELEMENTARY STRUCTURAL DRAWING.

PREFACE.

EXPERIENCE in teaching shows, that correct conceptions of the forms of objects having three dimensions, are obtained with considerable difficulty by the beginner, from drawings having but two dimensions, especially when those drawings are neither "natural"—that is "pictorial"—nor shaded, so as to suggest their form; but are artificial, or "conventional," and are merely "skeleton," or unshaded, line drawings. Hence moderate experience suggests, and continued experience confirms, the propriety of interposing, between the easily understood drawings of problems involving two dimensions, and the *general* course of problems of three dimensions, a *rudimentary* course upon the methods of representing objects having three dimensions. This much may be said in defence of Division I. of the present volume.

Experience again proves, in respect to the drawing of any engineering structures that are worth drawing, that it is a great advantage to the draftsman to have—1st, *some knowledge of the thing to be drawn*, aside from his knowledge of the methods of drawing it; and 2d, *practice* in the *leisurely study* of the graphical construction of single members or elements of a piece of framing, or other structure.

The truth of the second of the preceding remarks, is further apparent, from the fact that in entering at once upon the drawing of whole structures, three evils ensue, viz.—1st, *Confusion of ideas*, arising from the mass of new objects (the many different parts of a structure) thrown upon the mind at once; 2d, *Loss of time*, owing to repetition of the same detail many times in

the same structure; and 3d, *Waste of drawings*, as well as of
time, through poor execution, which is due to insufficient previous practice.

Hence Division II. is filled up with a tolerably liberal and
comprehensive collection of details of structures, each one of
which affords a useful problem, illustrative of some one feature
of *constructive art* and of *graphical procedure.*

Here it may be said, that the department of instruction, part
of which is embodied in this work, has involved in it a sub-
department, viz., that of *Descriptive Constructions, i.e.* a popular
descriptive study of the ways in which *many common things are
made and put together, or in which they operate.*

The classification of Chap. II., Div. II., is believed to be a
happy one, as readily suggesting various forms of framings.

Since my object is not so much to acquaint the student solely
with the works of the authors or artizans of other nations, as to
give them just those specimens of American methods of construction which are not usually found in books, I must refer
them to foreign works for examples of foreign modes of construction—not that such examples should be disparaged, but
because it is intended that this work shall *add* something to our
stock of book knowledge of the subjects of which it treats,
rather than merely repeat what has been before published.

Division III. contains a short special study of Shadows, taken
up in the form of the study of a few individual problems of a
practical character, assuming that the light comes only in the
simple conventional direction, commonly agreed upon by draftsmen, and saying nothing of the possibility of allowing the light
to be taken in *any* other direction.   The problems of this Division are intended to be sufficient, when fully understood, to
enable the student to construct accurately, or by a sufficiently
near approximation, all the shadows he will meet with in
Geometrical Drawing, till he shall pursue a general course on
shades and shadows.

Division IV. contains a sufficient exposition of the few and simple principles of *Isometrical projection* and *drawing*, to meet all the ordinary wants of the student or draftsman.

Division V. includes examples of a few simple structures, to fulfil the threefold purpose of affording occasion for learning the names of parts of structures; for practice in the combination of details into whole structures; and for profitable review practice in execution.

Finally, since experience has dictated so much of this preface, it shall dictate also these closing remarks. First: Classes will generally be found to take a lively interest in the subjects of this volume—because of their freshness to most learners, as new subjects of interesting study—because of the variety and brevity of the topics, which causes frequent change of the subject of study (such change being thought quite desirable by most learners)—and because of the compactness, neatness, and beauty of the volume which is formed by binding together all the plates of the course, when they are well executed. Second: As to the use of this volume, it is intended that there should be formal recitations of the problems in the 1st, 3d, and 4th divisions, with graphical constructions of a selection of the same or similar ones ; and occasional interrogations mingled with the graphical constructions of the practical problems of the remaining divisions. Remembering that excellence in mere execution, though highly desirable and to be encouraged, is not, at this stage of the student's progress, the sole end to be attained, a small number of graphic exercises will be sufficient. The student may, in place of a tedious course of finished drawings, be called on frequently to describe, by the aid of pencil or blackboard sketches, *how* he would construct drawings of certain objects—either those given in the several Divisions of this volume, or other similar ones proposed by his teacher.

In a course of school education, the full delivery to the student of the *greatest* number of *distinct ideas*, with the *least* amount of merely *manual labor*, should be the teacher's general aim.

A MANUAL

OF

ELEMENTARY GEOMETRICAL DRAWING.

DIVISION FIRST.
PROJECTIONS.

CHAPTER I.

FIRST PRINCIPLES.

§ I. *The purely Geometrical or Rational Theory of Projections.*

1. It is always easy to make a graphic solution of any problem involving but two dimensions, or a graphical construction of any object having but two dimensions, upon a plane surface, as a sheet of paper ; since, upon a surface having *two* dimensions, any figure having but two dimensions can have both of those dimensions represented in their true size and relative position.

2. The preceding article leads to the conclusion, that upon a plane surface, which has but *two* dimensions, no more than two dimensions can be represented in their true size and relative position. But when we consider the vast number of geometrical magnitudes, as cylinders, cones, spheres, &c., which have three dimensions, it appears probable, and actually is true, that the study of those magnitudes gives rise to a great many *problems* involving three dimensions. It appears, then, that there are problems of three as well as of two dimensions, and also that more than one plane surface is necessary to enable us to represent, both completely and accurately, bodies having three dimensions.

3. What then is the *number* and the *relative position* of the planes which will enable us to represent all the dimensions of any *geometrical solid*, in their real size, on those planes ? To assist in answering this question, reference may be made to Pl. I. Fig. 1. Let ABCFED be a regular square-cornered block, whose length is AB ; breadth, AD ; and thickness, AC ; and let MN be any horizontal

plane below it and parallel to its top surface ABED. If now the
four vertical edges of the block, at A, B, E, and D, be produced to
meet this plane, they will meet it in points—as a, b, e, and d. By
joining these points, it is evident that a figure—$abed$—will be
formed, which will be equal to the top surface of the block, and will
be a correct representation of the *length* and *breadth* of that top
surface—*i. e.* of the length and breadth of the block. Similarly,
let MP be a vertical plane, behind the block and parallel to its front
face, ABCF. Produce the four horizontal edges, through A, B, C,
and F, till they meet the vertical plane, MP, and connect the points,
a', b', c', and f', thus found. The figure, $a'c'b'f'$, so found, will
evidently be equal to the front face ABCF of the block, and will
therefore correctly represent the *length* and *thickness* of the block.

4. From the foregoing article, the following fundamental principles are deduced. *First :* Two planes, at right angles to each
other, are necessary to enable us to represent, fully, the three dimensions of a solid. *Second :* In order that those dimensions shall be
seen in their true size and relative position, they must be parallel to
that plane on which they are shown. *Third :* Each plane shows
two of the dimensions of the solid ; viz., the two which are parallel
to it, and that dimension which is thus shown twice, is the one
which is parallel to both of the planes. Thus AB, the length, and
AD, the breadth, are shown on the plane MN ; and AB, the length
again, and AC, the thickness, are shown on the plane MP.

5. From Art. (3), the following definitions naturally arise. If a
body be thrown directly forward, as across a table, all its points will
describe parallel straight lines, while the body will be said to be projected. Hence, as the lines of each group which are drawn through
the edges of the block, Pl. I., Fig. 1, to produce the figures $abed$
and $a'b'c'f'$, are parallel, the block is naturally said to be *projected*
upon the planes MN and MP; and the figures $abed$ and $a'b'c'f'$
are called the *projections* of the block. The planes, containing
these figures, are hence naturally called *planes of projection*—MN,
being horizontal, is called the *horizontal plane of projection*, and
MP, being vertical, is called the *vertical plane of projection*.

The projections take their particular names from the planes in
which they are found. Thus, $abed$ is the *horizontal projection* of the
block, and $a'b'c'f'$ is its *vertical projection*. The same idea is
expressed otherwise, by saying, that the block is *horizontally
projected* in the figure $abed$, and that it is *vertically projected* in the
figure $a'b'c'f'$.

The system of lines, of which Aa is one, may be called *horizon-*

tally projecting lines, and the group of which A*a'* is one, may be called *vertically projecting lines*. Finally: the line MR, in which the planes of projection intersect each other, is called the *ground line*.

6. The preceding definitions enable us to express a fourth fundamental principle, more briefly than could otherwise be done; viz. The height of the *vertical projection* of a point *above the ground line*, is equal to the *height of the point itself, in space, above the horizontal plane ;* and the perpendicular distance of the *horizontal projection* of a point *from the ground line*, is equal to the *perpendicular distance of the point itself, in front of the vertical plane*. Thus; Pl. I., Fig. 1, $aa'' = Aa'$ and $a'a'' = Aa$.

7. These four fundamental principles, with the accompanying definitions, are the foundation of the subject of projections, but, by attending carefully to Pl. I., Fig. 2, some additional special principles may be discovered, which are frequently applied in practice. Pl. I., Fig. 2, is a pictorial model of a pyramid, V*cdeg*, and of its two projections. The face, V*cd*, of the pyramid, is parallel to the vertical plane, and the triangle, X*ab*, is equal and parallel to V*cd*, and a little in front and at one side of it. By first conceiving, now, of the actual models, which are, perhaps, represented as clearly as they can be by mere diagrams, in Pl. I., Figs. 1 and 2; and then by attentive study of those figures, the facts stated in the next two articles will be fully comprehended.

§ II.— *Of the Relations of Lines to their Projections.*

8. *Relations of single lines to their projections.*

a. A *vertical line,* as AC, Pl. I., Fig. 1, has, for its horizontal projection, a point, *a*, and for its vertical projection, a line *a'c'*, perpendicular to the ground line, and equal and parallel to the line AC', in space.

b. A *horizontal line*, as AC, which is *perpendicular to the vertical plane*, has, for its horizontal projection, a line, *ad*, perpendicular to the ground line, and equal and parallel to the line, AD, in space ; and for its vertical projection a point, *a'*.

c. A *horizontal line*, as AB, which is *parallel to both planes* of projection, has, for *both* of its projections, lines *ab* and *a'b'*, which are parallel to the ground line, and equal and parallel to the line, AB, in space.

d. A *horizontal line*, as BD, which *makes an acute angle with the vertical plane*, has, for its horizontal projection, a line, *bd*, which makes the same angle with the ground line that the line, BD,

makes with the vertical plane, and is equal and parallel to the line
itself (BD); and has for its vertical projection a line $b'a'$, which is
parallel to the ground line, but shorter than BD, the line in space.

e. An *oblique line*, as BC, Pl. I., Fig. 1, or Vd, Pl. I., Fig. 2,
which is *parallel to the vertical plane*, has, for its vertical projection,
a line $b'c'$, or $v'd'$, which is equal and parallel to itself, and for its
horizontal projection, a line ba or vd, parallel to the ground line,
but shorter than the line in space.

f. An *oblique line*, as Vg, Pl. I., Fig. 2, *which is oblique to both
planes of projection*, has both of its projections, $v'd'$ and vg, oblique
to the ground line, and shorter than the line itself.

g. An *oblique line*, as AH, Pl. I., Fig. 1, which is oblique to both
planes of projection, but is *in a plane* ACDH, perpendicular to
both of those planes, has both of its projections, $a'c'$ and ad, perpendicular to the ground line, and shorter than the line itself.

h. A line, lying in either plane of projection, coincides with its
projection on that plane, and has its other projection in the ground
line. See cd—$c'd'$, the projections of cd, Pl. I., Fig. 2.

9. *Remark.* A general principle, which it is important to be
perfectly familiar with, is embodied in several of the preceding
examples; viz. When any line is *parallel to either plane of projection*, its projection on *that* plane is equal and parallel to itself, and
its projection *on the other plane* is parallel to the ground line.

10. The preceding remark serves to show how to find the true
length of a line, when its projections are given. When the line, as
Vg, Pl. I., Fig. 2, is oblique to both planes of projection, its length,
Vg, is evidently equal to the *hypothenuse* of a right-angled triangle,
of which the *base* is vg, the horizontal projection of the line, and the
altitude is Vv, the height of the upper extremity, V, above the
horizontal plane. When the line, as AH, Pl. I., Fig. 1, does not
touch either plane of projection, it is evidently equal to the hypothenuse of a right-angled triangle, of which the base, CH, equals
the *horizontal projection*, ad, and the altitude, AC, equals the
difference of *the perpendiculars*, Aa and Hd, to the horizontal plane.
In the same way, it is also true that the line, as Vg, Pl. I., Fig. 2,
is the hypothenuse of another right-angled triangle, whose base
equals the vertical projection, $v'd'$, and whose altitude equals the
difference of the perpendiculars, Vv' and $d'g$, from the extremities
of the line to the vertical plane of projection.

11. *Relations* of *pairs* of *lines to their projections.* These relations, after the full notice now given of the various positions of
single lines, may be briefly expressed as follows.

a. A pair of lines which are *equal and parallel in space, and also parallel to a plane* of *projection*, as AB and CF, Pl. I., Fig. 1, or V*c* and X*a*, Pl. I., Fig. 2, have their projections on that plane—*a′b′* and *c′f′*, Pl. I., Fig. 1, or *v′c′* and *x′a′*, Pl. I., Fig. 2—*equal and parallel—to each other, and to the lines in space.*

b. A pair of lines which are *equal and parallel* in space, *but not parallel to a plane* of projection, will have their projections on that plane equal and parallel to each other, but not to the lines in space.

c. Parallel lines *make equal angles* with either plane of projection ; hence it is easy to see that lines not parallel to each other— as V*d* and V*c*, or V*g* and V*e*, Pl. I., Fig. 2—but which make equal angles with the planes of projection, will have equal projections on both planes—i.e. $v′d′ = v′c′$ and $vg = ve$, also $vd = vc$.

§ III.—*Physical Theory of Projections.*

12. The preceding articles comprise the substance of the purely geometrical or rational theory of projections, which, strictly, is sufficient ; but it is natural to take account of the physical fact that the magnitudes in space and their representations, both address themselves to the eye, and to inquire *from what distance* and *in what direction* the magnitudes in Pl. I., Figs. 1 and 2, must be viewed, in order that they shall appear just as their projections represent them. Since the vertically-projecting perpendiculars, Q, regarded as rays of light, reflected from the block, Fig. 1, could only meet in the eye at an infinite distance in front of the vertical plane, we conclude that the vertical projection of an object represents that object, in respect to form, as it would appear to the eye situated at an infinite distance from it, and looking in a direction perpendicular to the vertical plane of projection. Reasoning similarly in reference to the group, S, we conclude that the horizontal projection of an object presents the same appearance, as respects form, that would be afforded by the object itself, when viewed by the eye at an infinite distance above it, looking in a direction perpendicular to the horizontal plane.

§ IV.—*Conventional Mode of representing the two Planes of Projection, and the two Projections of any Object upon one plane —viz. the Plane of the Paper.*

13. In practice, a single flat sheet of paper represents the two planes of projection, and in the following manner. The vertical plane, MV, Pl. I., Fig. 3, is supposed to revolve backwards, as

shown by the arcs ru and Vt, till it coincides with the horizontal plane produced at M u t G. Hence, drawing a line from right to left across the paper, to represent the ground line, MG, all that part of the paper above or beyond such a line will represent the vertical plane of projection, and the part below it the horizontal plane of projection.

14. Elementary Rational Geometry teaches that two lines, as PP', Pp, Pl. I., Fig. 3, which intersect, fix the position of the plane containing them. It also teaches that if from a point in space, as P, two lines P—P', P—p, be drawn—one and the other perpendicular to one and the other of two planes, MV and MH, which are at right angles to each other—then the plane, PP' $p''p$, of these lines, will be perpendicular to the two given planes, and will intersect those planes, in lines P'—p'' and p''—p, which will be perpendicular to the intersection, MG, of the planes, at the same point, p''.

This conclusion being reached, it can now be expressed more briefly by saying, that the *plane of the projecting lines* of any point, intersects the *planes of projection, in lines which are perpendicular to the ground line at the same point.*

15. Now, when the vertical plane revolves about the line MG, as an axis, the point p'', being in the axis, must remain fixed, and any other point, as P', must revolve in a plane perpendicular to MG. Hence the line P'—p'' will revolve—with the plane MV—to the position p'—p'', still remaining perpendicular to the ground line. Hence we have this fundamental principle of graphic constructions, viz. *The two projections p, p', Pl. I., Figs. 3 or 4, of a point in space, are always in a line, pp', perpendicular to the ground line.*

§ V.—*Of the Conventional Direction of the Light; and of the Position and Use of Heavy Lines.*

16. A single topic from DIVISION III. must here be anticipated, viz., that which refers to the assumption of a certain direction for the light. Without going into the subject fully, it will be sufficient to say, that as one stands facing the vertical plane of projection, the light is assumed to come from behind, and over the left shoulder, in such a direction that each of its projections makes an angle of 45° with the ground line, as shown in Pl. I., Fig. 6.

17. The *practical effect* of the preceding assumption in reference to the light, is, that upon a body of the form and position shown in Pl. I., Fig. 5, for example, the top, front, and left hand surfaces— *i.e.* the three seen in the Fig.—are illuminated, while the other three faces of the body are in the shade.

18. The *practical rule* by which the direction of the light and its effect are indicated in the projections, is, that all those *visible edges* of the body in space, which divide the light from the dark surfaces, are made heavy in projection.

19. To illustrate : The edges BC and CD of the body in space, Pl. I., Fig. 5, divide light from dark surfaces, and are seen in looking towards the vertical plane, and hence are made heavy in vertical projection, as seen at *b′c′* and *c′d′*. BK and KF divide illuminated from dark surfaces, and are seen in looking towards the horizontal plane, and are therefore made heavy in the horizontal projection, as shown at *bk* and *kf*.

20. By inspection, it will be seen that the following simple rule in reference to the position of the heavy lines on the drawings, may be deduced, as an aid to the memory. In all ordinary four-sided prismatic bodies, placed with their edges respectively parallel and perpendicular to the planes of projection, *or nearly so*, the *right hand lines, and those nearest the ground line, of both projections, are made heavy.*

21. Heavy lines are of considerable use, in the case of line drawings particularly, in indicating the forms of bodies, as will be seen in future examples.

§ VI.—*Notation.*

22. Under the head of Notation, two points are to be considered, the manner of indicating the various lines of the diagram, and the lettering. As will be seen by examining Pl. I., Figs. 1, 2 —see V*e, eg*, &c.—and 5, the *visible* lines of the object represented are indicated by *full* lines ; lines of construction and *invisible* lines of the object, so far as they are shown, are made in dotted lines. The intersections of auxiliary planes with the planes of projection, called *traces*, are represented by *broken and dotted* lines, as at P′QP″, Pl. I., Fig. 16.

23. Unaccented letters are used to indicate points of the horizontal projection; and the same with a single accent, denote the vertical projections of the same points. A strict adherence to the simple rule of lettering the same point with the same letter, wherever it is shown, will afford a complete key to very complicated diagrams, as will be shown as the course proceeds.

24. A little matter, which pertains to what might be called verbal notation, is, that in problems of a decidedly practical character, the term "*plan*" is generally used instead of "horizontal projection," and the term "*elevation*" is substituted for "vertical projection."

§ VII.—*Of the Use of the Method of Projections.*

25. Under this head it is to be noticed, that all drawings are made to serve one or the other of two purposes, *i.e.* they are made for *use* in aiding workmen in the construction of works; or in rendering intelligible, by means of drawings, the *real form and size* of some existing structure; or else, they are made for *ornament*, or to embellish our houses and gratify our tastes, and to show the *apparent forms and relative sizes* of objects.

26. Drawings of the former kind are often called, on account of the uses to which they are applied, "*mechanical*" or "*working*" *drawings.* Those of the latter kind are commonly called pictures; and here it is to be noticed that if "working" drawings are to show the *true*, and not the *apparent*, proportions of all parts of an object, they must, all and always, conform to this one rule, viz. *All those lines which are equal and similarly situated on the object, must be equal and similarly situated on the drawing.*

But, as is now abundantly evident, drawings made according to the method of projections, do conform to this rule; hence their use, as above described.

CHAPTER II.

27. *Remark.*—The style of *execution* of the problems of this chapter is so simple, and so nearly alike for all of them, that it need not be described for each problem separately, but will be noticed from time to time in connection with groups of figures which are similarly executed.

§ I.—*Projections of Straight Lines.*

28. PROB. 1. *To construct the projections of a vertical straight line,* 1½ *inches long, whose lowest point is* ½ *an inch from the horizontal plane, and all of whose points are* ¾ *of an inch from the vertical plane.*

Remarks. a. The remaining figures of Pl. I. are drawn just half the size indicated by the dimensions given in the text. It may be well for the student to make them of full size.

b. Let MG be understood to be the ground line for all of the above problems, without further mention of it.

1st. Draw, very lightly, an indefinite line perpendicular to the ground line, Pl. I., Fig. 7.

2d. Upon it mark a point, a', two inches above the ground line, and another point, b', half an inch above the ground line.

3d. Upon the same line, mark the point a,b, three-fourths of an inch below the ground line. Then $a' b'$ will be the vertical, and ab the horizontal projection of the required line. (8 *a*)

29. PROB. 2. *To construct the projections of a horizontal line,* 1½ *inches long,* 1½ *inches above the horizontal plane, perpendicular to the vertical plane, and with its furthermost point—from the eye* —¼ *of an inch from that plane.* Pl. I., Fig. 8, in connection with the full description of the preceding problem, will afford a sufficient explanation of this one.

Remark. It often happens that a diagram is made more intelligible by lettering it as at *ab*, Pl. I., Fig. 7, and at *c'd'*, Pl. I. Fig. 8, for thus the notation shows unmistakably, that *ab* or *c'd'* are not the projections of points but of lines.

30. PROBLEMS 3 to 8, inclusive, need now only to be enunciated, with references to their constructions, in Pl. I.

Fig. 9 shows the projections of a line, $2\frac{1}{4}$ inches long, parallel to the ground line, $1\frac{1}{2}$ inches from the horizontal plane, and 1 inch from the vertical plane.

Fig. 10 is the representation of a line, 2 inches long ; parallel to the horizontal plane, and 1 inch above it ; and making an angle of 30° with the vertical plane.

Fig. 11 represents a line, $2\frac{1}{4}$ inches long, parallel to the vertical plane, and $1\frac{3}{4}$ inches from it, and making an angle of 60° with the horizontal plane.

Fig. 12 gives the projections of a line, $1\frac{1}{2}$ inches long, lying in the horizontal plane, parallel to the ground line, and $1\frac{1}{4}$ inches from it.

Fig. 13 shows the projections of a line, $1\frac{1}{4}$ inches long, lying in the vertical plane, parallel to the ground line, and 1 inch from it.

Fig. 14 indicates a line, $2\frac{1}{2}$ inches long, lying in the vertical plane, and making an angle of 60° with the horizontal plane.

31. PROB. 9. *To construct the projections of a line which is in a plane perpendicular to both planes of projection, the line being oblique to both planes of projection.* Pl. I., Fig. 15, shows a pictorial model of this problem. AB represents the line in space ; *ab* its horizontal projection ; *a'b'* its projection on the principal vertical plane M P'; and A'B' its projection on an auxiliary vertical plane P'QP'' which is parallel to the given line AB. Since the line AB is parallel to this auxiliary plane, its projection upon it— A' B'—equals A B.

32. Now in causing all three of the planes just referred to, to coincide with the plane of the paper, taken as a horizontal plane, the plane P'QP'' is revolved about P'Q, as an axis, till it coincides with the vertical plane produced, as at P'QV''; and then the whole vertical plane MP'V'' is revolved backwards into the horizontal plane of projection. Now observe, with regard to the first of these revolutions, that as the axis P'Q is a vertical line—being the intersection of two vertical planes—all points of the arc A'*a''* will be at the same height above the horizontal plane MP''. The same is true of the arc B'*b''*, hence *a''* and *a'* are at equal heights above

DIV. 1.

PL. 1.

the ground line, and so are b'' and b', from which facts we may pass at once to the general principle that *two or more different elevations* (vertical projections) (24) *of the same point, as* A, *will be in a line,* $a'a''$, *parallel to the ground line,* or in space, at equal heights, $a't$ and $A'a''''$ above the common horizontal plane MP''.

33. Some other principles may be noted, and remarks made, in connection with Pl. I., Fig. 15.

a. $A'm$ being the vertically-projecting line of the point A', ma'' is the vertical projection of the arc $A'a''$ on the plane QV''. $a'''a''''$ is evidently the horizontal projection of the same arc. A moment's reflection upon these facts will lead at once to the principle, that *when an arc, or circle, is parallel to either plane of projection, its projection on that plane will be an equal arc, or circle, and on the other plane, a straight line parallel to the ground line.*

b. $a''''b''''$ may be called the horizontal projection of the projection of AB on the plane $P'QP$, and it evidently coincides with the projection of ab on the same plane. Likewise, mn may be called the vertical projection of the projection of AB on the plane $P'QP$, and it is plainly identical with the projection of the projection $a'b'$ upon the same plane $P'QP$.

c. Inspection of the figure shows that the distance Bb' of any point, as B, from the vertical plane of projection equals bt, and that bt equals $b''n$, the distance of the second elevation of B from the vertical trace $P'Q$ of the auxiliary vertical plane.

d. Further inspection shows, that the second elevation shows the line in its true length, that the angle made by $a''b''$ with $H'q$, equals the angle made by the line AB with the horizontal plane, and that the angle made by $a''b''$ with $P'Q$ equals the angle made by AB with the vertical plane. Furthermore, each of these angles is evidently the complement of the other.

34. After this extended preliminary examination of all the facts connected with this problem, as shown in Pl. I., Fig. 15, every point in Pl. I., Fig. 16, will probably be intelligible, as like parts of both figures have the same letters. Let the line be 2 inches long, and let it make an angle of 60° with the horizontal plane, $a''b''$ is the auxiliary elevation on the vertical plane $P'QP$, after that plane has been revolved into the primitive vertical plane, and shows the true length and position of the given line. Suppose the line in space to be $1\frac{1}{2}$ inches to the left of the auxiliary vertical plane $P'QP$ then $a'b'$, its vertical projection, will be perpendicular to the ground line, between the parallels $a''a'$ and $b''b'$ (32), and $1\frac{1}{2}$ inches from $P'Q$. The horizontal projection, ab, will be in $a'b'$

produced. $b''n$—$b'''b''''$ are the two projections of the arc in which
the point b'' revolves back to its position, n—b'''', in the plane
P'QP, and $b''''b$—nb' is the line in which nb'''' is projected back
to its primitive position $b'b$. Therefore, b is at the intersection
of $b''''b$ with $a'b'$ produced. a is similarly found, giving ab as the
horizontal projection of the given line.

35. *Execution.* The foregoing problems are to be inked with
very black ink; the projections of given lines, and the ground line,
in *heavy full* lines; and the lines of construction in *fine dotted* lines
as shown in the figures. Lettering is not necessary, except for
purposes of reference, as in a text book, though it affords occasion
for practice in making small letters.

On the other hand, poorly executed lettering disfigures a
diagram so much that it should be made only after some previous
practice, and then carefully; making the letters *small*, *fine*, and
regular.

§ II.—*Right Projections of Solids.*

36. *Remark.* The term " right projection " becomes significant
only when it refers to bodies which are, to a considerable extent,
bounded by straight lines at right angles to each other. Such
bodies are said to be drawn in right projection when most of their
bounding lines are parallel or perpendicular to one or the other
of the planes of projection.

37. PROB. 10.—*To construct the projections of a vertical right
prism, having a square base ; standing upon the horizontal plane,
and with one of its faces parallel to the vertical plane.* Pl. II.,
Fig. 17.

Let the prism be 1 inch square, 1½ inches high, and ⅛ of an inch
from the vertical plane.

1*st.* The square ABEF, ⅛ of an inch from the ground line, is the
plan of the prism, and strictly represents its upper base.

2*d.* A'B'C'D', 1½ inches high, is the elevation of the prism, and
strictly represents its front face.

3*d.* The vertical edge A'C', is horizontally projected at A; and
so each corner of the plan is the horizontal projection of some one
of the vertical *edges* of the prism.

4*th.* The vertical face A'B'C'D' has for its horizontal projection
the line AB; and so each side of the plan is the horizontal projection of one of the vertical *faces* of the prism.

5*th.* The horizontal edge AE, is vertically projected at A'; and so each corner of the elevation is the vertical projection of some edge which is perpendicular to the vertical plane.

6th. The vertical projection of the upper base, is A′B′; the vertical projection of the lower base, is C′D′, because that lower base is in the horizontal plane of projection.

7th. The right hand vertical face of the prism, has for its vertical projection B′D′, and for its horizontal projection BF.

The style of *execution* of the figures of this plate is, in general, sufficiently indicated by the diagrams.

38. PROB. 11.—*To construct the plan and two elevations of a prism having the proportions of a brick, and placed with its length parallel to the ground line.* Plate II., Fig. 18.

1st. abcd is the plan, ¾ of an inch broad, twice that distance in length, and ⅜ of an inch from the ground line, showing that the prism in space is at the same distance from the vertical plane of projection.

2nd. a′b′e′f′ is the elevation, ⅜ of an inch thick, and as long as the plan; and ⅞ of an inch above the ground line, showing that the prism in space is at this height above the horizontal plane.

3rd. If a plane, P′QP, be placed perpendicular to both of the principal planes of projection, and touching the right hand end of the prism, it is evident that the projection of the prism upon such a plane will be a rectangle, equal, in length, to the width, *bd*, of the plan, and, in height, to the height, *b′f′* of the side elevation. This new projection will also, evidently, be at a distance from the primitive vertical plane, *i.e.* from P′Q, equal to *d*Q, and at a distance from the horizontal plane equal to Q*f′*. When, therefore, the auxiliary plane, P′QP, is revolved about P′Q into the primitive vertical plane of projection, the new projection will appear at *a″e″c″g″*.

4th. dc‴ is the horizontal, and *b′c″* the vertical projection of the arc in which the point *db′* revolves into the primitive vertical plane. *ba‴*, *b′a″*, are the two projections of the horizontal arc in which the corner *bb′* of the prism revolves.

39. PROB. 12.—*To construct the two projections of a cylinder which stands upon the horizontal plane.* Pl. II., Fig. 19.

The circle AaB*b* is evidently the plan of such a cylinder, and the rectangle A′B′C′D its elevation. Observe, here, that while the elevation, alone, is the same as that of a prism of the same height, Fig. 17, the plan shows the body represented, to be a cylinder.

40. As regards execution, the right hand line B′D′ of a cylinder

or cone may be made less heavy than the line B′D′, Fig. 17 ; and
in the plan, the semicircle, *a*B*b*, convex towards the ground line,
and limited by a diameter *ab*, which makes an angle of 45° with
the ground line, is made heavy, but gradually tapered, into a fine
line in the vicinity of the points *a* and *b*.

41. PROB. 13.—*To construct the projections of a cylinder whose
axis is placed parallel to the ground line.* Pl. II., Fig. 20.

Let the cylinder be 1½ inches long, ¾ of an inch in diameter, its
axis ¾ of an inch from the horizontal plane, and ½ an inch from the
vertical plane. The principal projections will, of course, be two
equal rectangles, *gehf* and *a′b′c′d′*, since all the diameters of
the cylinder are equal. The centre lines, *g′h′* and *ab*, are made at
the same distances from the ground line, that the axis of the cylinder is from the planes of projection.

42. Taking now into consideration the mere *form* of these two
projections, they could not be distinguished from the projections of a
square-based prism of equal length and diameter. There are, however, three ways of distinguishing between the projections of a horizontal cylinder and of a horizontal prism having a square base.
First: In the case of the cylinder, the lines *c′d′* and *ef*, are made
quite moderately heavy, rather than fully heavy, as they would be
on the projections of a prism. *Second:* The lettering shows plainly,
at a glance, that Pl. II., Fig. 20, represents a cylinder ; thus, *g′h′* is
the vertical projection of three lines ; viz., the element *gh*, which is
furthest from the vertical plane, *ab*, considered as the horizontal
projection of the axis, and *ef*, the element nearest to the vertical
plane. Again, *ab* is the horizontal projection of three lines ; viz., of
a′b′ the highest element, of *g′h′* the axis, and of *c′d′*, the lowest
element. *Third:* An end elevation shows, that in Pl. II., Fig. 20,
a cylinder is represented. In the figure, the auxiliary vertical plane
P′QP does not coincide with the base of the cylinder, but is at a distance, *ef′′′*, from it, and is revolved towards the left, to coincide
with the primitive vertical plane of projection. Excepting this,
the construction is so similar to that of Prob. 11, as to need no
further explanation.

In *inking*, the end elevation, *b′′f′′d′′*, is made heavy at *nf′′d′′p*,
and tapered into a fine line in the vicinity of *n* and *p*.

§ III.— *Oblique Projections showing two sides of a Solid Right
Angle.*

43. A *solid* right angle is an angle such as that at any corner of

a cube, or like the space in the corner of a square room, and is bounded by the three *plane* right angles which form such corners.

When a cube, for example, is held so that two of its vertical faces can be seen, we should not look at them directly or in a direction perpendicular to either of them, and the cube would be said to be viewed obliquely, and where drawn in this position, would be said to be represented in oblique projection.

44. PROB. 14.—*To construct the plan and two elevations of a vertical prism, with a square base; resting on the horizontal plane, and having its vertical faces inclined to the vertical plane of projection.* Pl. II., Figs. 21–22.

1*st.* ABCG is the plan, with its sides placed at any convenient angle with the ground line.

2*d.* A'B'D'E' is the vertical projection of that vertical face whose horizontal projection is AB.

3*d.* B'C'E'F' is the vertical projection of that face whose horizontal projection is BC. This completes the vertical projection of the visible parts of the prism, when we look at the prism in the direction of the lines CF', &c.

4*th.* Let *gb* be the horizontal trace of an auxiliary vertical plane of projection, which is perpendicular to both of the principal planes of projection. In looking perpendicularly towards this plane, *i.e.* in the directions G*g*, &c., AG and AB are evidently the horizontal projections of those vertical faces that would then be visible; and the projecting lines, G*g*, A*a*, and B*b* determine the widths *ga* and *ab* of those faces as seen in the new elevation. Now the auxiliary plane *gb* is not necessarily revolved about its vertical trace (not shown), but may just as well be taken up and *transferred* to any position where it will coincide with the primitive vertical plane; only its ground line *gb* must be made to coincide with the principal ground line, as at H'E". Hence, making H"D" and D"E" respectively equal to *ga* and *ab*, and by drawing H"G", &c., the new elevation will be completed.

45. By inspecting the two elevations—Pl. II., Figs. 21–22—it appears that they are identical in form, but that the parts of Fig. 22 are not identical with those of Fig. 21.

To prevent the student from constructing this problem merely by rote, the same faces on the two elevations are distinguished by marks. Thus the surfaces marked ♯ are the two elevations of the same face of the prism; the one marked φ is visible only on the first elevation, and the one marked × is visible only on the second elevation—Fig. 22.

46. Pl. II., Fig. 23, represents a small quadrangular prism in two elevations, the axis being horizontal in space, so that the left hand elevation shows the base of the prism. In the practical applications of this construction, the centre, *s*, of the square projection is generally on a given line, not parallel to the sides of the square. Hence this construction affords occasion for an application of the problem: *To draw a square of given size, with its centre on a given line, and its sides not parallel to that line.* The following solution should be carefully remembered, it being of frequent application. Through the given centre, *s*, draw a line, L, in any direction, and another, L′, also through *s*, at right angles to L. On *each* of these lines, lay off *each way* from *s*, half the length of a side of the square. Through the points thus formed, draw lines *parallel* to the lines L and L′ and they will form the required square whose centre is *s*.

47. PROB. 15.—*To construct the plan and several elevations of a vertical hexagonal prism, which rests upon the horizontal plane of projection.* Pl. II., Figs. 24, 25, 26.

Remark. The distinction between right and oblique projection becomes obscure as we pass from the consideration of bodies whose surfaces are at right angles to each other. Figs. 24 and 25 show a hexagonal prism as much in right projection as such a body can be thus shown, but, as in both cases a majority of its surfaces are, considered separately, in oblique projection, its construction is given here.

48. In Fig. 24 the hexagonal prism is, as shown by the plan, placed so that two of its vertical faces are parallel to the vertical plane of projection. Observe that where the hexagon is thus placed, three of its faces will be visible, one of them in its real size, viz,. BC, B′C′F′G′, and that the extreme width, E′H′, of the elevation, equals the diameter of the circumscribing circle of the plan. Notice, also, that as BC equals half of AD, while AB and CD are equal, and equally inclined to the vertical plane, the elevations, A′F′ and G′D′, of these latter faces, *will be equal, and each half as wide as the middle face.* This fact enables us to construct the elevation of a hexagonal prism situated as here described, without constructing the plan, provided we know the width and height of one face of the prism. This last construction should be remembered, it being of frequent and convenient application in the drawing of nuts, bolt-heads, &c., in machine drawing.

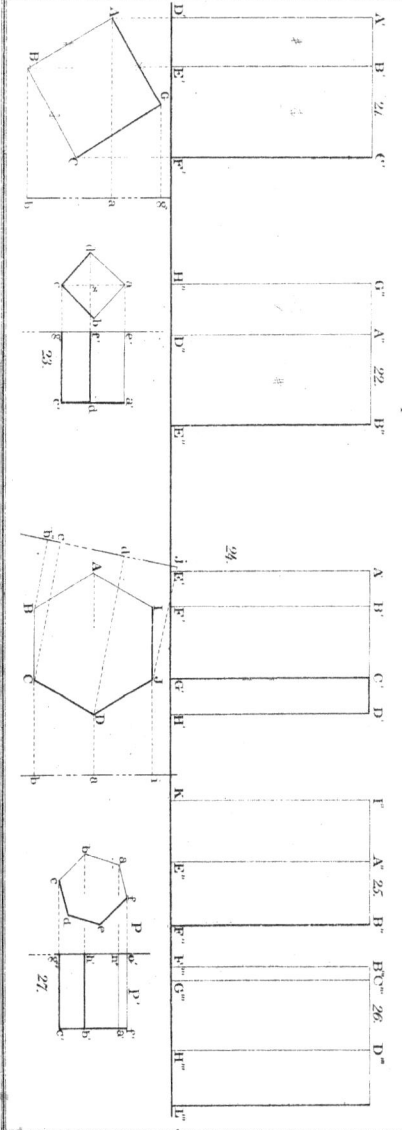

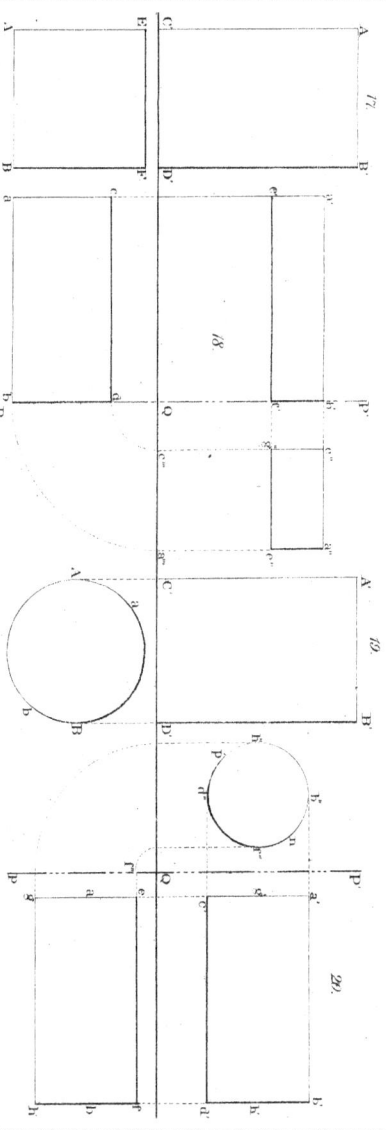

49. Pl. II., Fig. 25 shows the elevation of the same prism on a plane which originally was placed at *ib*, and perpendicular to the horizontal plane ; whence it appears, that if a certain elevation of a hexagonal prism shows three of its faces, and one of them in its full size, another elevation, at right angles to this one, will show but two faces, neither of them in its full size ; the extreme width, I″B″, of the second elevation being equal to the diameter of the inscribed circle of the plan.

50. Pl. II., Fig. 26 *shows the elevation of the same prism as it would appear* if projected upon a vertical plane having the position indicated by the horizontal trace *jb″*, when that plane had been transferred to the principal vertical plane at Fig. 26. In this elevation, none of the faces of the prism are seen in their true size. The auxiliary vertical plane, whose horizontal trace is *jb″*, could have been revolved about that trace, directly back into the horizontal plane, causing the corresponding elevation to appear in the lines D*d*, &c., produced to the left of *jb″* as a ground line.

Elevations on auxiliary vertical planes can always be made to appear thus, but it seems more natural to have the several elevations side by side above the principal ground line—a result which is accomplished by transferring the auxiliary planes as heretofore described. Fig. 27 represents two elevations of a hexagonal prism, placed so as to show the base in one elevation, and three of its faces, unequally, in the other. The centre of the elevation which shows the base, may be made in a given line perpendicular to *o′g′*, by placing the centre of the circumscribing circle used in constructing the hexagon, upon such a line. Having constructed this elevation, project its points, *a,b*, &c., across to the other vertical plane, P′, which is in space perpendicular to the plane, P, at the line, *o′g′*. By representing the elevation on P′ as touching *o′g′*, we indicate that the prism touches the plane, P, just as the elevation in Fig. 24, indicates that the prism there shown rests upon the horizontal plane.

51. PROB. 16.—*To construct the plan and two elevations of a pile of blocks of equal widths, but of different lengths, so placed as to form a symmetrical body of uniform width.* Pl. III., Fig. 28.

The auxiliary vertical plane of projection, perpendicular to the horizontal plane at *k‴f‴′*, is made to coincide with the principal vertical plane by direct revolution. The point *a‴a″″*, the projection of aa' on the auxiliary vertical plane, revolves in a horizontal arc, of which $a'''a''$ is the horizontal, and $a''''a''$ the vertical pro-jection (33 a), giving a'', a point of the second elevation. Other points of this elevation are found in the same way. This figure differs from Figs. 18 and 20, of Plate II., only in presenting more points to be constructed. If the student finds any difficulty with this example, let him refer to those just mentioned, and to first principles.

52. PROB. 17. *To construct the oblique projection of a vertical circle.* Pl. III., Fig. 30.

Let the circle first be placed parallel to the vertical plane of pro-jection, and tangent to the horizontal plane. Its vertical projection will then be a circle $a'b'e'g'$, tangent to the ground line, and its horizontal projection will be a straight line, parallel to the ground line, and equal to the diameter of the circle. Now let a vertical line, F—F'f', be drawn tangent to the circle at FF', and let the circle be revolved about this line as an axis—still being kept ver-tical—till it makes a certain angle with the vertical plane, as shown by the angle which the new horizontal projection, FB, makes with the ground line. Now, in this revolution, any point, $b'b$, of the primitive position of the circle, describes a horizontal arc in space, of which bB is the horizontal projection—equal to the arc in space (33 a)—and b'B' is the vertical projection, giving B' a point in the new elevation. Other points, as $c'c$, $a'a$, which, being in a vertical line, are horizontally projected in the same point, c—a, revolve in horizontal arcs, of which c—a; A—C is the horizontal projection, and a'A' c'C'; are the vertical projections, giving A' and C', other points of the new elevation. The remaining points of the new elevation are found in a manner precisely similar, as is fully shown in the figure. This being a plane problem, no part of it need be inked in heavy lines.

53. PROB. 18.—*To construct the projections of a cylinder whose convex surface rests on the horizontal plane, and whose axis is inclined to the vertical plane.* Pl. III., Fig. 31.

As may be learned by reflecting upon Fig. 19, of Plate II., the principle is general that the projection of a right cylinder with a circular base, upon any plane to which its axis is parallel, will be a rectangle. Therefore let CSTV, Pl. III., Fig. 31, be the plan of the cylinder. Since it rests upon the horizontal plane, $q'u'$, in the ground line, is the vertical projection of its line of contact with that plane, and $p'A'$ is the vertical projection of pA, the highest element of the cylinder, as it is at a height above the ground line, equal to the diameter, TV, of the cylinder. The construction of the vertical projection of either base, is but a repetition of the last problem. In the figure, the left hand base is found just as in the preceding problem, and the construction, being fully given, needs no further explanation.

54. The vertical projection of the base TV is found somewhat differently. TV may properly be considered as the horizontal diameter of this base. Then let the base be revolved about TV, till it becomes parallel to the horizontal plane of projection. It will then appear as a circle, and a line, as $n''n$, will show the true height of n above the diameter TV. So, also, $o''o$ will show the true distance of o below TV. Therefore the vertical projections of the points n and o, will be in the line n—n', perpendicular to the ground line, and at distances above and below TV', the vertical projection of TV, equal, respectively, to nn'' and oo''. Having, in the same manner, found r' and t', the vertical projections of two points whose common horizontal projection t—r is assumed, as was n—o, the vertical projection of the base TV can be drawn by the help of the irregular curved ruler.

55. In the *execution* of this figure, SV is made slightly heavy, and TV fully heavy, and, for reasons which cannot here be fully explained, the portion, $n'T't'$, of the elevation of the right hand base, and the small portion, $D'u'$, of the left hand base, are made heavy. Suffice it to say: *First*. That a part of the convex surface is in the light, while the right hand base is in the dark. *Second*. $n'T't'$ divides the illuminated half of the convex surface, from the base at the right, which is in the dark ; and $D'u'$ divides the illuminated left hand base from the visible portion of the darkened half of the convex surface (18–20).

56. Prob. 19. *To construct the two projections of a right cone, with a circular base in the horizontal plane ; and to construct either projection of a line, drawn from the vertex to the circumference of the base, having the other projection of the same line given.* Pl. III., Fig. 32.

Remark. When the *axis* of a cone is vertical, perpendicular to the vertical plane, or parallel to the ground line, the cone is shown in right projection as much as such a body can be, but as all the straight lines upon its surface are then inclined to one or both planes of projection, the above problem is inserted here among problems of oblique projections.

57. Let VB be the radius of the circle, which, with the point V, is the horizontal projection of the cone. Since the base of the cone rests in the horizontal plane of projection, C'B' is its vertical projection. Since the axis of the cone is vertical, V', the vertical projection of the vertex, must be in a perpendicular to the ground line, through V, and may be assumed, unless the height of the cone is given. V'C' and V'B', the extreme elements, as seen in elevation, are parallel to the vertical plane of projection, hence their horizontal projections are CV and BV, parallel to the ground line (8 e). Let it be required to find the horizontal projection of any element, whose vertical projection, V'D', is given. V is the horizontal projection of V', and D', being in the circumference of the base, is horizontally projected at D, therefore VD is the horizontal projection of that straight line in the conic surface, whose vertical projection is V'D'. Having given, VA, the horizontal projection of a straight line in the surface of the cone, let it be required to find its vertical projection. V' is the vertical projection of V, and A, being in the circumference of the base, is vertically projected at A'. Therefore V'A' is the required vertical projection of the proposed line. In inking the figure, no part of the plan is heavy lined, and in the elevation, only the element V'B' is slightly heavy.

58. PROB. 20. *To construct the projections of a right hexagonal prism ; whose axis is oblique to the horizontal plane, and parallel to the vertical plane.* Pl. III., Figs. 33, 34.

1st. Commence by constructing the projections of the same prism as seen when standing vertically, as in Fig. 33. The plan only is strictly needed, but the elevation may as well be added here, for completeness' sake, and because some use can be made of it.

2nd. Draw J''G'', making any convenient angle with the ground line, and set off upon it spaces equal to G'J', J'H', and J'I', from Fig. 33.

3rd. Since the prism is a right one, at J'', &c., draw perpendiculars to J''G'', make each of them equal to J''C', Fig. 33, and draw F''C'', which will be parallel to J''G'', and will complete the second elevation.

4th. Let us suppose that the prism was moved from its first position, Fig. 33, parallel to the vertical plane, and towards the right, and then inclined, as described, with the corner, CJ', of the base, remaining in the horizontal plane. It is clear that all points of the new plan, as B''', would be in parallels, as BB''', to the ground line, through the primitive plans, as B, of the same points. It is equally true that the points of the new plan will be in perpen-diculars to the ground line through the new elevations B″, &c., of the same points (15), hence these points B‴, &c., will be at the intersections of these two groups of lines. Thus, A‴ is at the intersection of AA‴ with A″A‴; C‴ is at the intersection of CC‴ with C″C‴; K‴ is at the intersection of DK‴ with H″K‴, &c.

5th. B‴C‴, F‴E‴, and G‴K‴, being the projections of lines of the prism which are parallel in space, are themselves parallel. A similar remark applies to C‴D‴, A‴F‴, and H‴G‴. Observe, that as the upper or visible base is viewed obliquely, it is not seen in its true size, F‴C‴ being less than FC, Fig. 33; so that this base A‴C‴, E‴, does not appear in the new plan as a regular hexagon.

59. PRÓB. 21. *To construct the projections of the prism, given in the previous problem, when its edges are inclined to both planes of projection.* Pl. III., Fig. 34a.

Remark. This problem involves the showing of three faces of a solid angle, but, with a similar one respecting the pyramid, is retained here, on account of its usefulness as a study, in giving completeness to the general problem of the projection of the prism.

60. If the prism, Pl. III., Fig. 34, be moved to any new position, such that the inclination of its edges to the vertical plane, only, shall be changed, the inclination of its edges to the horizontal plane of projection being unchanged, the new plan will be merely a copy of the second plan, placed in a new position. Let the par-ticular position chosen be such that the axis of the prism shall be in a plane perpendicular to the ground line, i.e. to both planes of projection; then the axis of symmetry, C‴G‴, of the second plan, will take the position C‴′G‴′, and on each side of this line the plan, Fig. 34 a, will be made, similar to the halves of the plan in Fig. 34.

As the prism is turned horizontally about the corner J″, and then transferred, producing the result that the inclination of its axis to the horizontal plane is unchanged, all points of the third elevation, as A‴′′, C‴′′, &c., will be in parallels to the ground line through A″, C″, &c., and in perpendiculars to the ground line, through A‴′, C‴′, &c.

61. By examination of this solution, and by inspection of Figs. 34 and 34a, it appears that a change in the position of the axis, with reference to but one plane of projection at a time, can be

represented directly from projections already given; also that a curve, beginning with the first plan, and traced through the six figures composing the three given pairs of projections in the order in which they *must* be made, would be an S curve, ending in the third elevation.

62. *Execution.*—The full explanation of the location of the heavy lines cannot here be given. The careful inquirer may be able to satisfy himself that the heavy lines of the figures, as shown, are the projections of those edges of the prism which divide its illuminated from its dark surfaces.

63. PROB. 22. *To construct the projections of a regular hexagonal pyramid, whose axis is inclined to the horizontal plane only.* Pl. III., Figs. 35, 36.

1st. Commence, as with the prism in the last problem, by representing the pyramid as having its axis vertical.

2nd. Draw $a''d''$, equal to $a'd'$, and divided in the same way. At n'', the middle point of $a''d''$, draw $n''V''$ perpendicular to $a''d''$, and make it equal to $n'V'$, which gives V'' the new elevation of the vertex. Join V'' with a'', b'', c'', and d'', and the new elevation will be completed.

3rd. Supposing the same translation and rotation to occur to the primitive position of the pyramid, that was made in the case of the prism (52, *4th*), the points of the new plan, Fig. 36, will be found in a manner similar to that shown in Fig. 34. V''' is at the intersection of VV''' with $V''V'''$; c''' is at the intersection cc''' with $c''c'''$; d''' is at the intersection of dd''' with $d''d'''$, &c.

4th. The points, $a'''b'''c'''$ . . . . f''', of the base, are connected with V''', the new horizontal projection of the vertex, to complete the new plan. If the pyramid were less inclined, the perpendicular $V''V'''$ would fall within the base, and the whole base would then be visible in the plan. As it is, $f'''a'''$ and $a'''b'''$ are hidden, and therefore dotted.

5th. The heavy lines are correctly placed in the diagram; also the partially heavy lines, which are all between $V'''d'''$ and the ground line, but the reasons for their location cannot here be given, beyond the general principle (18–20) already given.

64. PROB. 23. *To construct the projections of the regular hexagonal pyramid, when its axis is oblique to both planes of projection.* Pl. III., Fig. 36a.

Suppose the pyramid here shown to be the one represented in

figures 35 and 36, and suppose that it has been turned horizontally about the corner, $a''a'''$, Fig. 36, of the base, and then transferred to the position represented in Fig. 36a. After the pyramid has been thus moved, the third plan will be merely a copy of the second plan, placed so that its axis of symmetry, $V''''d''''$, shall make any assumed angle with the ground line. Make the distances on $V''''d''''$ equal to those on $V'''d'''$, and let $c''''e''''$ and $b''''f''''$ equal $c'''e'''$ and $b'''f'''$, after which the third plan is completed by joining the points of the base with V'''', the vertex.

Next we have to consider, that as the inclination of the axis of the pyramid to the horizontal plane of projection is unchanged, the points, as V''''', of the third elevation, will be at the intersection of parallels to the ground line, through the corresponding points, as V'', of the second elevation, with perpendiculars through the same points, as V'''', seen in the third plan. Observe that the two points vertically projected in c'', being at the same height above the ground line, will appear in the third elevation at c''''' and e''''', in the same straight line, through c'', and parallel to the ground line.

Remembering also that lines which are parallel in space must have parallel projections, on the same plane, $c'''''d'''''$ will be parallel to $f'''''a'''''$, &c.

The heavy lines are indicated in the figure.

DIVISION SECOND.

DETAILS OF CONSTRUCTIONS IN MASONRY, WOOD, AND METAL.

CHAPTER I.

CONSTRUCTIONS IN MASONRY.

§ I.—*General Definitions and Principles applicable both to Brick and Stone-work.*

65. A horizontal layer of brick, or stone, is called a *course*. The seam between two courses is called a *coursing-joint*. The seam between two stones or bricks of the same course, is a *vertical* or *heading-joint*. The vertical joints in any course should abut against the solid stone or brick of the next courses above and below. This arrangement is called *breaking joints*. The particular arrangement of the pieces in a wall is called its bond. As far as possible, stones and bricks should be laid with their *broadest surfaces horizontal*. Bricks or stones, whose length is in the direction of the length of a wall, are called *stretchers*. Those whose length is in the direction of the thickness of a wall, are called *headers*.

§ II.—*Brick Work.*

66. If it is remembered that bricks used in building have, usually, an invariable size, $8'' \times 4'' \times 2''$ (the accents indicate inches), and that in all ordinary cases they are used whole, it will be seen that brick walls can only be of certain thicknesses, while, in the use of stone, the wall can be made of any thickness.

Thus, to begin with the thinnest house wall which ever occurs, viz. one whose thickness equals the length of a brick, or 8 inches; the next size, disregarding for the present the thickness of mortar, would be the length of a brick added to the width of one, or equal to the width of three bricks, making 12 inches, a thickness employed in the partition walls and upper stories of first class houses, or the

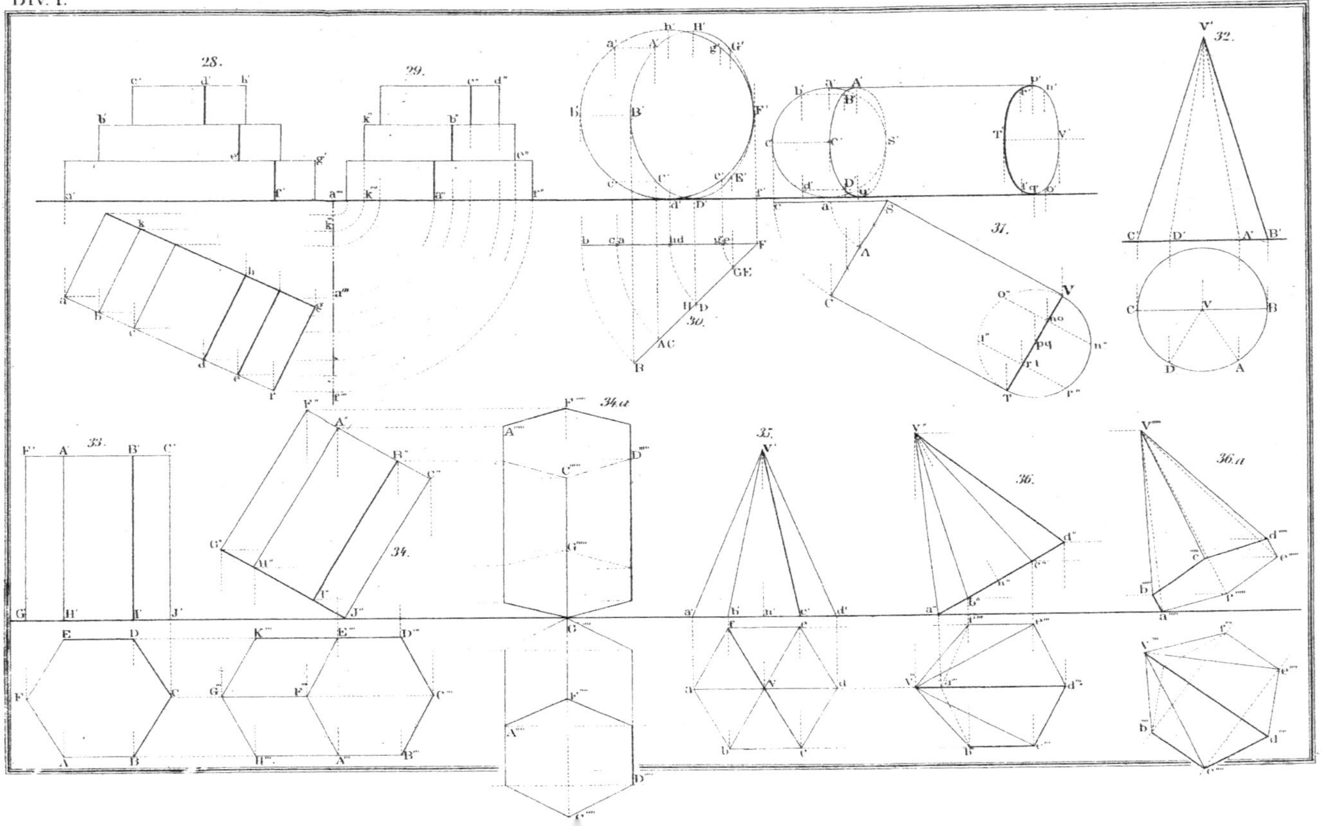
28.
29.
30.
31.
32.
33.
34.
34.a
35.
36.
36.a

outside walls of small houses. Then, a wall whose thickness is equal to the length of two bricks or the width of four, making 16 inches, a thickness proper for the outside walls of the lower stories of first class houses; and lastly, a wall whose thickness equals the length of two bricks added to the width of one; or, equals the width of five bricks, or 20 inches, a thickness proper for the basement walls of first class houses, for the lower stories of few-storied, heavy manufactory buildings, &c.

67. In the common bond, generally used in this country, it may be observed—

a. That in heavy buildings a common rule appears to be, to have one row of headers in every six or eight rows of bricks or courses, i.e. five or seven rows of stretchers between each two successive rows of headers; and,

b. That in the 12 and 20 inch walls there may conveniently be a row of headers in the back of the wall, intermediate between the rows of headers in the face of the wall, while in the 8 inch and 16 inch walls, the single row of headers in the former case, and the double row of headers in the latter, would take up the whole thickness of the wall, and there might be no intermediate rows of headers.

c. The separate rows, making up the thickness of the wall in any one layer of stretchers, are made to break joints in a horizontal direction, by inserting in every second row a half brick at the end of the wall.

68. Calling the preceding arrangements common bonds, let us next consider the bonds used in the strongest engineering works which are executed in brick. These are the *English bond* and the *Flemish bond*.

The English Bond.—In this form of bond, every second course, as seen in the face of the wall, is composed wholly of headers, the intermediate courses being composed entirely of stretchers. Hence, in any practical case, we have given the thickness of the wall and the arrangement of the bricks in the front row of each course, and are required to fill out the thickness of the wall to the best advantage.

The Flemish Bond.—In this bond, each single course consists of alternate headers and stretchers. The centre of a header, in any course, is over the centre of a stretcher in the course next above or below. The face of the wall being thus designed, it remains, as before, to fill out its thickness suitably.

69. EXAMPLE 1. **To represent an Eight Inch Wall in English Bond.** Let each course of stretchers consist of two rows, side

by side, the bricks in which, break joints with each other horizontally. Then the joints in the courses of headers, will be distant half the width of a brick from the vertical joints in the adjacent courses of stretchers, as may be at once seen on constructing a diagram.

70. Ex. 2. **To represent a Twelve Inch Wall in English Bond.** See Pl. IV., Fig. 37. In the elevation, four courses are shown. The upper plan represents the topmost course, and in the lower plan, the second course from the top is shown. The courses having stretchers in the face of the wall, could not be filled out by two additional rows of stretchers, as such an arrangement would cause an unbroken joint along the line, *ab*, throughout the whole height of the wall—since the courses having headers in the face, *must* be filled out with a single row of stretchers, in order to make a twelve inch wall, as shown in the lower plan.

In order to allow the headers of any course to break joints with the stretchers of *the same* course, the row of headers may be filled out by a brick, and a half brick—split lengthwise—as in the upper plan ; or by two three-quarters of bricks, as seen in the lower plan.

71. Ex. 3. **To represent a Sixteen Inch Wall in English Bond.** The simplest plan, in which the joints would overlap properly, seems to be, to have every second course composed entirely of headers, breaking joints horizontally, and to have the intermediate courses composed of a single row of stretchers in the front and back, with a row of headers in the middle, which would break joints with the headers of the first named courses. If the stretcher courses were composed of nothing but stretchers, there would evidently be an unbroken joint in the middle of the wall extending through its whole height.

72. Ex. 4. **To represent an Eight Inch Wall in Flemish Bond.** Pl. IV., Fig. 38, shows an elevation of four courses, and the plans of two consecutive courses. The general arrangement of both courses is the same, only a brick, as AA', in one of them, is set six inches to one side of the corresponding brick, B, of the next course —measuring from centre to centre.

73. Ex. 5. **To represent a Twelve Inch Wall in Flemish Bond.** Pl. IV., Fig. 39, is arranged in general like the preceding figures, with an elevation, and two plans. One course being arranged as indicated by the lower plan, the next course may be made up in two ways, as shown in the upper plan, where the grouping shown at the right, obviates the use of half bricks in every second course.

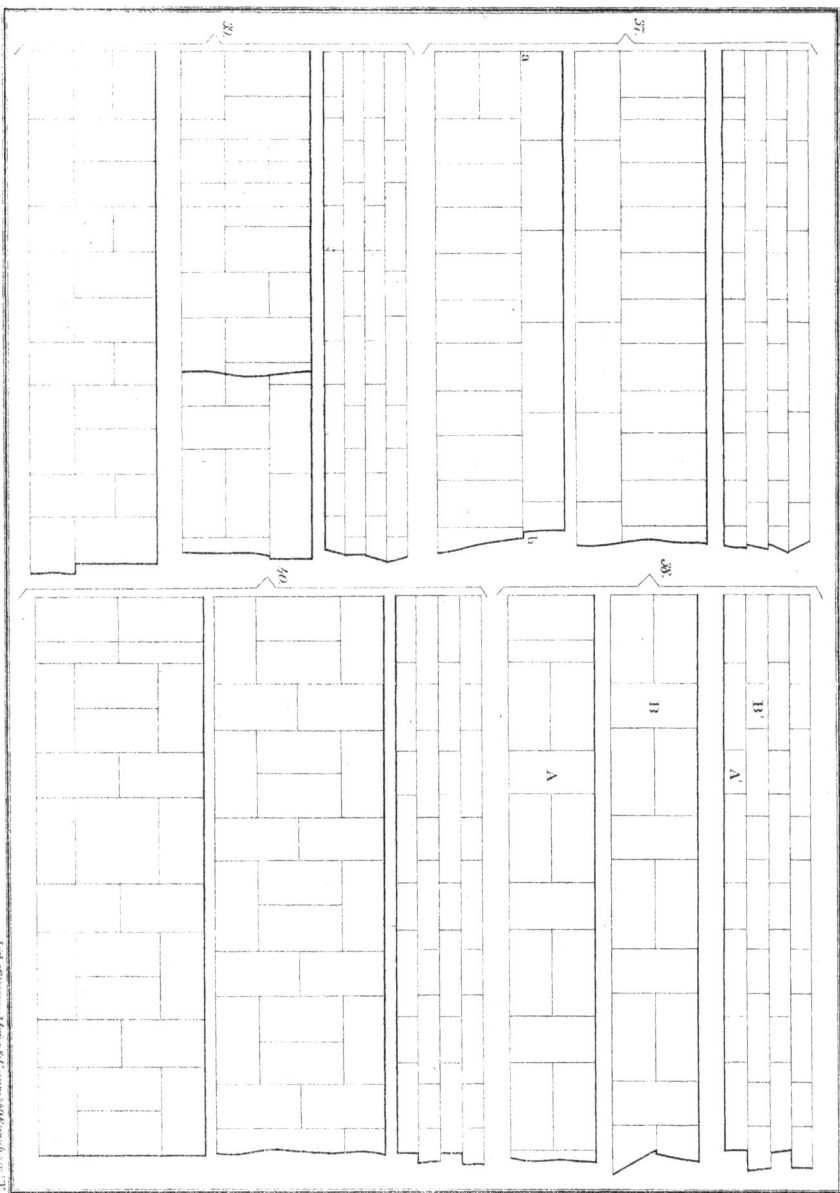

Pl. IV.

There seems to be no other simple way of combining the bricks in this wall so as to avoid the use of half bricks, without leaving open spaces in some parts of the courses.

74. Ex. 6. **To represent a Sixteen Inch Wall in Flemish Bond.** Pl. IV., Fig. 40. The figure explains itself sufficiently. Bricks may not only be split crosswise and lengthwise, but even thicknesswise, or so as to give a piece $8 \times 4 \times 1$ inches in size. Although, as has been remarked, whole bricks of the usual dimensions can only form walls of certain sizes, yet, by inserting fragments, of proper sizes, any length of wall, as between windows and doors, or width of pilasters or panels, may be, and often is, constructed. By a similar artifice, and also by a skilful disposition of the mortar in the vertical joints, tapering structures, as tall chimneys, are formed.

§ III.—*Stone Work.*

75. The following examples will exhibit the leading varieties of arrangement of stones in walls.

EXAMPLE 1. **Regular Bond in Dressed Stone.** Pl. V. Fig. 41. Here the stones are laid in regular courses, and so that the middle of a stone in one course, abuts against a vertical joint in the course above and the course below. In the present example, those stones whose ends appear in the front face of the wall, seen in elevation, take up the whole thickness of the wall as seen in plan.

The right hand end of the wall is represented as broken down in all the figures of this plate. Broken stone is represented by a smooth broken line, and the under edge of the outhanging part of any stone, as at *n*, is made heavy.

76. Ex. 2. **Irregular Rectangular Bond.** Pl. V., Fig. 42. In this example, each stone has a rectangular face in the front of the wall. These faces are, however, rectangles of various sizes and proportions, but arranged with their longest edges horizontal, and also so as to break joints.

77. That horizontal line of the plan which is nearest to the lower border of the plate, is evidently the plan of the top line of the elevation, hence all the extremities, as a', b', &c., of vertical joints, found on that line, must be horizontally projected as at a and b, in the horizontal projection of the same line.

78. Ex. 3. **Rubble Walls.** The remaining figures of Pl. V., represent various forms of " rubble " wall. Fig. 43 .represents a wall of broken boulders, or loose stones of all sizes, such as are found abundantly in New England. Since, of course, such stones

would not fit together exactly, the "chinks" between them are filled with small fragments, as shown in the figure. Still smaller irregularities in the joints, which are not thus filled, are represented after tinting by heavy strokes in inking. Fig. 44 represents the plan and elevation of a rubble wall made of slate; hence, in the plan, the stones appear broad, and in the elevation, long and thin, with chink stones of similar shape. Fig. 45 represents a rubble wall, built in regular courses, which gives a pleasing effect, particularly if the wall have cut stone corners, of equal thickness with the rubble courses.

79. *Execution.*—The graphical *construction* of objects being now more an object of study than the mere style of *execution* employed, the graining of the stone work on Pl. V. is omitted.

Plate V. may be, 1*st*, pencilled; 2*d*, inked in fine lines; 3*d*, tinted. The rubble walls, having coarser lines for the joints, may better be tinted, before lining the joints in ink.

The plans need not be tinted. ·

Also, in case of the rubble walls, sudden heavy strokes may be made occasionally in the joints, to indicate slight irregularities in their thickness, as has already been mentioned.

The right hand and lower side of any stone, not joining another stone on those sides, is inked heavy, in elevation, and on the plans as usual. The left hand lines of Figs. 43 and 44 are tangent at various points to a vertical straight line, walls, such as are represented in those figures, being made vertical, at the finished end, by a plumb line, against which the stones rest.

The shaded elevations on Pl. V., may serve as guides to the *depth* of color to be used in tinting stone work. The tint actually to be used, should be composed of Prussian blue, with a little India ink, and carmine; building stone, for engineering purposes, being usually bluish gray granite, or limestone.

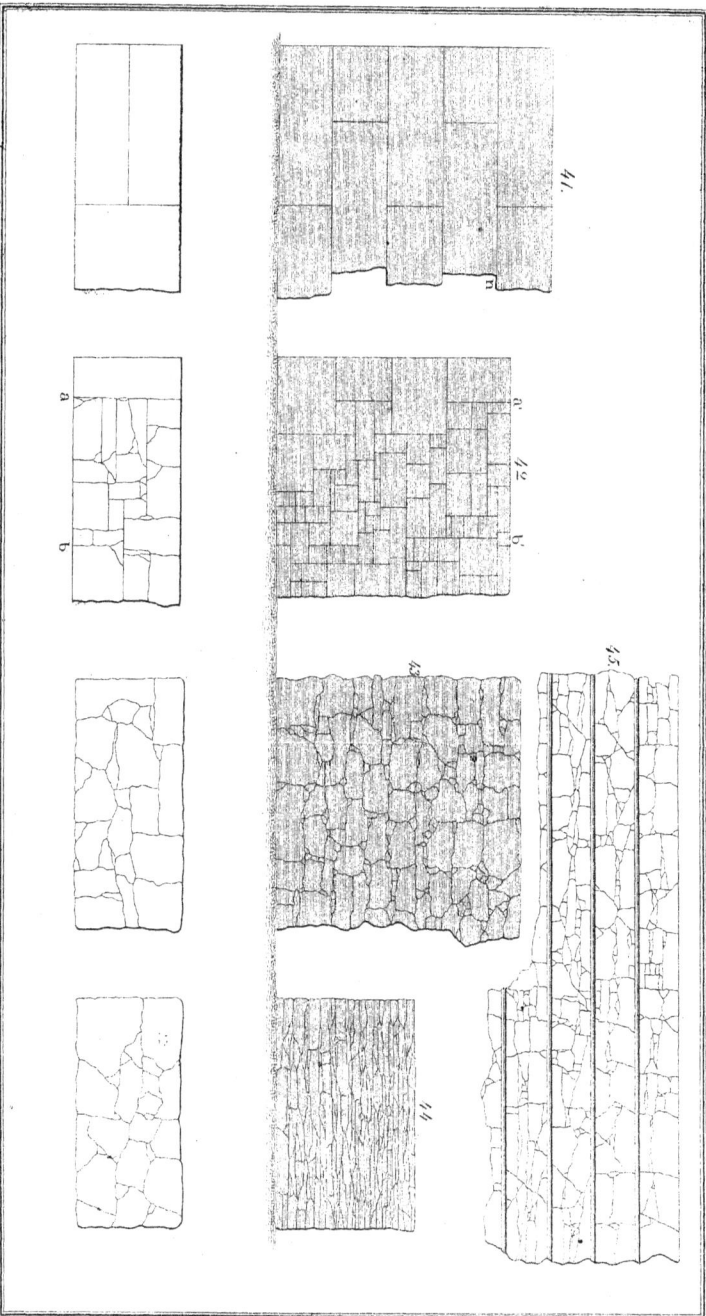

CHAPTER II.

§ I.—*General Remarks.*

80. Two or more beams may be framed together, so as to make any angle with each other, from 0° to 180°; and so that the plane of two united pieces may be vertical, horizontal, or oblique.

81. To make the present graphical study of framings more fully rational, it may here be added, in respect to each of the positions of a pair of pieces, named in the last article, that they may be framed with reference to the resisting of forces which would act to separate them in the direction of any one of the three dimensions of each. Following out, in regular order, the two-fold classification involved in this and in the preceding article, let us presently proceed to notice several examples—some only by popular description of their *material construction* and *action*, and some by a complete description of their *graphical construction* and *execution*, also.

82. Two other points may here be, however, further introduced, as further grounds for a systematic treatment of the present subject. *First:* A pair of pieces may be *immediately* framed into each other, or they may be *intermediately* framed by " bolts," " keys," &c., or both modes may be, and often are, combined. *Second:* Two combinations of timbers which are alike in general appearance, may be adapted, the one to resist extension, and the other, compression, and may have slight corresponding differences of construction.

83. *Note.*—For the benefit of those who may not have had access to the subject, the following brief explanation of scales, &c., is here inserted.

Drawings, showing the pieces as taken apart so as to show the mode of union of the pieces represented, are called " *Details.*"

The drawings about to be described, are to be made in plan, side and end elevations, sections and details, or in as few of these views as will show clearly all parts of the object represented.

84. In respect to the instrumental operations, these drawings are

supposed to be "made to scale," from measurements of models, or from assumed measurements.　It will, therefore, be necessary, before beginning the drawings, to explain the manner of sketching the object, and of taking and recording its measurements.

85. In sketching the object, make the sketches in the same way in which they are to be drawn, i.e. *in plan and elevation*, and not in perspective, and make enough of them to contain all the measurements, i.e. to show all parts of the object.

In measuring, take measurements of all the parts which are to be shown; and not merely of individual parts alone, but such connecting measurements as will locate one part with reference to another.

86. The usual mode of recording the measurements, is, to indicate, by arrow heads, the extremities of the line of which the figures between the arrow heads show the length.

87. For brevity, an accent (') denotes feet, and two accents (") denote inches.　The dimensions of small rectangular pieces are indicated as in Pl. VI., Fig. 50, and those of small circular pieces, as in Fig. 51.

88. In the case of a model of an ordinary house framing, such as it is useful to have in the drawing room, and in which the sill is represented by a piece whose section is about $2\frac{1}{2}$ inches by 3 inches, a scale of one inch to six inches is convenient.　Let us then describe this scale, which may also be called a scale of two inches to the foot.

The same scale may also be expressed as a scale of one foot to two inches, meaning that one foot on the object *is represented by* two inches on the drawing ; also, as a scale of $\frac{1}{6}$, thus, a foot being equal to twelve inches, 12 inches on the object is represented by two inches on the drawing ; therefore, *one* inch on the drawing represents *six* inches on the object, or, *each* line of the drawing is $\frac{1}{6}$ of the same line, as seen upon the object ; each *line*, for we know from Geometry that *surfaces* are to each other as the squares of their homologous dimensions, so that if the length of the *lines* of the drawing is *one-sixth* of the length of the same lines on the object, the *area* of the drawing would be *one thirty-sixth* of the area of the object, but the scale always refers to the relative *lengths* of the *lines* only.

89. In constructing the scale above mentioned, upon the stretched drawing paper,

1*st.* Set off upon a fine straight pencil line, two inches, say three times, making four points of division.

2*d.* Number the left hand one of these points, 12, the next, 0, the next, 1, the next, 2, &c., for additional points.

3*d.* Since each of these spaces represents a foot, if any one of them, as the left hand one, be divided into twelve equal parts, those parts will be representative inches. Let the left hand space, from (12) to (0) be thus divided, by fine vertical dashes, into twelve equal parts, making the three, six, and nine inch marks longer, so as to catch the eye, when using the scale.

4*th.* As some of the dimensions of the object to be drawn are measured to quarter inches, divide the first and sixth of the inches, already found, into quarters; dividing two of them, so that each may be a check upon the other, and so that there need be no continual use of one of them, so as to wear out the scale.

5*th.* When complete, the scale may be inked; the length of it in fine parallel lines about $\frac{1}{20}$ of an inch apart.

90. It is now to be remarked that these spaces are always to be called by the names of the dimensions they represent, and not according to their actual sizes, i. e. the space from 1 to 2 represents a foot upon the object, and is called a foot; so each twelfth of the foot from 12 to 0 is called an inch, since it represents an inch on the object; and so of the quarter inches.

91. Next, is to be noticed the directions in which the feet and inches are to be estimated.

The feet are estimated from the zero point towards the right, and the inches from the same point towards the left.

Thus, to take off 2′—5″ from the scale, place one leg of the dividers at 2, and extend the other to the fifth inch mark beyond 0, to the left; or, if the scale were constructed on the edge of a piece of card-board, the scale being laid upon the paper, and with its graduated edge against the indefinite straight line on which the given measurement is to be laid off, place the 2′ or the 5″ mark, at that point on the line, from which the measurement is to be laid off, according as the given distance is to be to the left or right of the given point, and then with a needle point mark the 5″ point or the 2′ point, respectively, which will, with the given point, include the required distance.

92. *Remark.* The preceding directions are somewhat circuitous and abstract, for want of an illustrative diagram, but are purposely so, for being exact, they will doubtless be intelligible, and after constructing the scale for himself as directed in Art. 89, the student will readily comprehend its use, as now explained.

93. Other scales, constructed and divided as above described,

only smaller, are found on the ivory scale, marked 30, &c., meaning 30 feet to the inch when the tenths at the left are taken as
feet; and meaning three feet to the inch when the larger spaces
—three of which make an inch—are called feet, and the twelfths
of the left hand space, inches. So, on the other side of the ivory,
are found scales marked $\frac{5}{8}$, &c., meaning scales of $\frac{5}{8}$ inch to *one*
foot, or *ten* feet, according as the whole left hand space, or its
tenth, is assumed as representing one foot. *Note* that $\frac{5}{8}$ of an inch
to a foot is $\frac{2}{3}$ of a foot to the inch, $\frac{5}{8}$ of an inch to *ten* feet, is 16
feet to an inch, &c.

94. Of the immense superiority of drawing by these scales, over
drawing without them, it is needless to say much: without them,
we should have to go through a mental calculation to find the
length of every line of the drawing. Thus, for the piece which is
two and a half inches high, and drawn to a scale of two inches to
a foot, we should say—2$\frac{1}{2}$ inches$=\frac{2\frac{1}{2}}{12}$ of a foot$=\frac{5}{24}$ of a foot.
One foot on the object$=$two inches on the drawing, then $\frac{1}{24}$ of a
foot on the object$=\frac{1}{24}$ of 2 inches$=\frac{2}{24}=\frac{1}{12}$ of an inch, and $\frac{5}{24}$ of a
foot $(=2\frac{1}{2}$ inches$)=\frac{5}{24}$ of 2 inches$=\frac{5}{12}$ of an inch.

A similar tedious calculation would have to be gone through
with for every dimension of the object, while, by the use of scales,
like that already described, we take off the same number of the feet
and inches of the scale, that there are of real feet and inches in any
given line of the object.

§ II.—*Pairs of Timbers whose axes make angles of 0° with each
other.*

95. EXAMPLE 1. A Compound Beam bolted, Pl. II., Fig. 46.
Mechanical Construction.

The figure represents one beam as laid on top of another. Thus
situated, the upper one may be slid upon the lower one in the
direction of two of its three dimensions; or it may rotate about
any one of its three dimensions as an axis. A single bolt, passing
through both beams, as shown in the figure, will prevent all of
these movements except rotation about the bolt as an axis. Two
or more bolts will prevent this latter, and consequently, all move-
ment of either of the beams upon the other. A bolt, it may be
necessary to say, is a rod of iron whose length is a little greater
than the aggregate thickness of the pieces which it fastens together. It is provided at one end with a solid head, and at the other,
with a few screw threads on which turns a " nut," for the purpose

of gradually compressing together the pieces through which the bolt passes.

96. *Graphical Construction.* Assuming for simplicity's sake in this and in most of these examples, that the timbers are a foot square, and having the given scale; the diagrams will generally explain themselves sufficiently. The scales are expressed fractionally, adjacent to the numbers of the diagrams. The nut only is shown in the plan of this figure.

It is an error to suppose that the nuts and other small parts can be carelessly drawn, as by hand, without injury to the drawing, since these parts easily catch the eye, and if distorted, or roughly drawn, appear very badly.

The method is, therefore, here fully given for drawing a nut accurately. Take any point in the centre line, *ab*, of the bolt, produced, and through it draw any two lines, *cd* and *en*, at right-angles to each other. From the centre, lay off *each way* on *each line*, half the length of each side of the nut, say $\frac{3}{4}$ of an inch.

Then, through the points so found, draw lines parallel to the centre lines *cd* and *en*, and they will form a square plan of a nut $1\frac{1}{2}''$ on each side.

In making this construction, the distances should be set off very carefully, and the sides of the nut *ruled*, in very fine lines, and exactly through the points located. From the plan, the elevation is found as in Pl. II., Fig. 21.

Note.—The Teacher need not make the same selection, from these examples, of specimens to be actually drawn by classes, that is here given; neither need he confine himself to *any* of those given in the present collection of exercises; in fact, wherever practicable, it would be well to take examples from real structures, using the specimens here given, as a guide in making the selections.

97. Ex. 2. **A Compound Beam, notched and bolted.** Pl. VI., Fig. 47. *Mechanical Construction.* The beams represented in this figure, are indented together by being alternately notched; the portions cut out of either beam being a foot apart, a foot in length, and two inches deep. When merely laid, one upon another, they will offer resistance only to being separated longitudinally, and to horizontal rotation.

The addition of a bolt renders the "compound beam," thus formed, capable of resisting forces tending to separate it in all ways.

Thin pieces are represented, in this figure, between the bolt-head and nut, and the wood. These are circular, having a rounded edge, and a circular aperture in the middle through which the bolt passes. They are called "*washers*," and their use is, to distribute the pressure of the nut or bolt-head over a larger surface, so as not to indent the wood, and so as to prevent a gouging of the wood in tightening the nut, which gouging would facilitate the decay of the wood, and consequently, the loosening of the nut.

98. *Graphical Construction.*—The beams being understood to be originally one foot square, the compound beam will be 22 inches deep; hence draw the upper and lower edges 22 inches apart, and from each of them, set off, on a vertical line, 10 inches. Through the points, *a* and *b*, so found, draw *very faint* horizontal lines, and on either of them, lay off any number of spaces; each, one foot in length. Through the points, as *c*, thus located, draw transverse lines between the faint lines, and then, to prevent mistakes in inking, make slightly heavier the notched line which forms the real joint between the timbers.

The use of the scale of $\frac{1}{24}$ continues till a new one is mentioned.

The following empirical rules will answer for determining the sizes of nuts and washers on assumed sketches like those of Pl. VI., so as to secure a good appearance to the diagram. The side of the nut may be double the diameter of the bolt, and the greater diameter of the washer may be equal to the diagonal of the nut, plus twice the thickness of the washer itself.

Execution.—This is manifest in this case, and in most of the following examples, from an inspection of the figures.

99. Ex. 3. **A Compound Beam, keyed.**   Pl. VI., Fig. 48. *Mechanical Construction.* The defect in the last construction is, that the bearing surfaces opposed to separation in the direction of the length of the beam, present only the ends of the grain to each other. These surfaces are therefore liable to be readily abraded or made spongy by the tendency to an interlacing action of the fibres. Hence it is better to adopt the construction given in Pl. VI., Fig. 48, where the "*keys*," as K, are supposed to be of hard wood, whose grain runs in the direction of the width of the beam. In this case, the bolts are passed through the keys, to prevent them from slipping out, though less boring would be required if they were placed midway between the keys. In this example, the strength of the beam is greatly increased with but a very small increase of material, as is proved in mechanics and confirmed by experiment.

100. *Graphical Construction.*—This example differs from the last so slightly as to render a particular explanation unnecessary. The

keys are 12 inches in height, and 6 inches in width, and are 18 inches apart from centre to centre. They are most accurately located by their vertical centre lines, as AA'. If located thus, and from the horizontal centre line BB', they can be completely drawn before drawing *ee'* and *nn'*. The latter lines, being then pencilled, only between the keys, mistakes in inking will be avoided.

Execution.—The keys present the end of their grain to view, hence are inked in diagonal shade lines, which, in order to render the illuminated edges of the keys more distinct, might terminate, uniformly, at a short distance from the upper and left hand edges.

By shading only that portion of the right hand edge of each key, which is between the timbers, it is shown that the keys do not project beyond the front faces of the timbers.

101. Ex. 4. **A Compound Beam, scarfed.** Pl. VI., Fig. 49

Mechanical Construction. This specimen shows the use of a *series of shallow notches* in giving one beam a firm hold, so to speak, upon another; as *one deep notch*, having a bearing surface equal to that of the four shown in the figure, would so far cut away the lower beam as to render it nearly useless.

102. *Graphical Construction.*—The notches, *one foot* long, and *two inches* deep, are laid down in a manner similar to that described under Ex. 2.

103. *Execution.*—The keys, since they present the end of the grain to view, are shaded as in the last figure. Heavy lines on their right hand and lower edges would indicate that they projected beyond the beam.

Remark.—When the surfaces of two or more timbers lie in the same plane, as in many of these examples, they are said to be "*flush*" with each other.

§ III.—*Combinations of Timbers, whose axes make angles of* 90° *with each other.*

104. The usual way of fastening timbers thus situated, is by means of a projecting piece on one of them, called a "*tenon*," which is inserted into a corresponding cavity in the other, called a "*mortise.*" The tenon may have three, two, or one of its sides flush with the sides of the timber to which it belongs; while the mortise may extend entirely, or only in part, through the timber in which it is made, and may be enclosed by that timber on three or on all sides. [See the examples which follow, in which some of these cases are represented, and from which the rest can be understood.]

When the mortise is surrounded on three or on two sides, particularly in the latter case, the framed pieces are said to be "*halved*" together, more especially in case they are of equal thickness, and have half the thickness of each cut away, as at Pl. VI., Fig. 52.

105. EXAMPLE 1. **Two examples of a Floor Joist and Sill.** (From a Model.) Pl. VI., Fig. 53. *Mechanical Construction.* A—A' is one sill, B—B' another. CC' is a floor timber framed into both of them. At the left hand end, it is merely " dropped in," with a tenon; at the right hand end, it is framed in, with a tenon and " tusk," e. At the right end, therefore, it cannot be *lifted* out, but must be *drawn* out of the mortise. The tusk, e, gives as great a thickness to be broken off, at the insertion into the sill, and as much horizontal bearing surface, as if it extended to the full depth of the tenon, t, above it, while less of the sill is cut away. Thus, labor and the strength of the sill, are saved.

106. *Graphical Construction.*—1*st*. Draw ab. 2*d*. On ab construct the elevation of the sills, each $2\frac{1}{2}$ inches by 3 inches. 3*d*. Make the two fragments of floor timber with their upper surfaces flush with the tops of the sills, and 2 inches deep. 4*th*. The mortise in A', is $\frac{3}{4}$ of an inch in length, by 1 inch in vertical depth. 5*th*. Divide cd into four equal parts, of which the tenon and tusk occupy the second and third. The tenon, t, is $\frac{3}{4}$ of an inch long, and the tusk, e, $\frac{1}{4}$ of an inch long. Let the scale of $\frac{1}{8}$ be used.

107. *Execution.*—The sills, appearing as sections in elevation, are shaded. In all figures like this, dotted lines of construction should be freely used to assist in " reading the drawing," i.e. in comprehending, from the drawing, the construction of the thing represented.

108. Ex. 2. **Example of a "Mortise and Tenon," and of "Halving."** (From a Model.) Pl. VI., Fig. 54. *Mechanical Construction.* In this case, the tenon, AA', extends entirely through the piece, CC', into which it is framed. B and C are halved together, by a mortise in each, whose depth equals half the thickness of B, as shown at B'' and C'', and by the dotted line, ab.

Graphical Construction.—Make, 1*st*, the elevation, A'; 2*d*, the plan; 3*d*, the details. B'' is an elevation of B as seen when looking in the direction, BA. C'' is an elevation of the left hand portion of CC', showing the mortise into which B is halved. The dimensions may be assumed, or found by a scale, as noticed below.

109. *Execution.*—The invisible parts of the framing, as the halv-

ing, as seen at *ab* in elevation, are shown in dotted lines. The brace and the dotted lines of construction serve to show what separate figures are comprehended under the general number (54) of the diagram. The scale is ⅛. From this the dimensions of the pieces can be found on a scale.

110. Ex. 3. **A Mortise and Tenon as seen in two sills and a post. Use of broken planes of section.** (From a Model.) Pl. VI., Fig. 55.

Mechanical Construction.—The sills, being liable to be drawn apart, are pinned at *a*. The post, BB′, is kept in its mortise, *bb″*, by its own weight; *m* is the mortise in which a vertical wall joist rests. It is shown again in section near *m′*.

111. *Graphical Construction.*—The plan, two elevations, and a broken section, show all parts fully.

The assemblage is supposed to be cut, as shown in the plan by the broken line AA′A″A‴, and is shown, thus cut, in the shaded figure, A′A′A‴*m′*. The scale, which is the same as in Fig. 53, indicates the measurements. At B″, is the side elevation of the model as seen in looking in the direction A′A.

In Fig. 55 *a*, A′s obviously equals A″s, as seen in the plan.

112. *Execution.*—In the shaded elevation, Fig. 55*a*, the cross-section, A′A‴, is lined as usual. The longitudinal sections are shaded by longitudinal shade lines. The plan of the broken upper end of the post, B, is filled with arrow heads, as a specimen of a way, sometimes convenient, of showing an end view of a broken end.

Sometimes, though it renders the execution more tedious, narrow blank spaces are left on shaded ends, opposite to the heavy lines, so as to indicate more plainly the situation of the illuminated edges (100). The shading to the left of A′, Fig. 55*a*, should be placed so as to distinguish its surface from that to the right of A′.

113. Ex. 4. **A Mortise and Tenon, as seen in timbers so framed that the axis of one shall, when produced, be a diagonal diameter of the other.** Pl. VI., Fig. 56. *Mechanical Construction.*—In this case the end of the inserted timber is not square, and in the receiving timber there is, besides the mortise, a tetraedron cut out of the body of that timber.

114. *Graphical Construction.*—D is the plan, D′ the side elevation, and D″ the end elevation of the piece bearing the tenon. F′ and F are an elevation and plan of the piece containing the mortise. Observe that the middle line of D, and of D′, is an axis of symmetry, and that the right hand edges of D and D′ are parallel to the sides of the incision in F′.

§ IV.—*Miscellaneous Combinations.*

115. EXAMPLE 1. **Dowelling.** (From a Model.) Pl. VI., Fig. 57. *Mechanical Construction.*—*Dowelling* is a mode of fastening by pins, projecting usually from an edge of one piece into corresponding cavities in another piece, as seen in the fastening of the parts of the head of a water tight cask. The mode of fastening, however, rather than the relative position of the pieces, gives the name to this mode of union.

The example shown in Pl. VI., Fig. 57, represents the braces of a roof framing as dowelled together with oak pins.

116. *Graphical Construction.*—This figure is, as its dimensions indicate, drawn from a model. The scale is one-third of an inch to an inch.

1*st*. Draw *acb*, with its edges making any angle with the imaginary ground line—not drawn.

2*d*. At the middle of this piece, draw the pin or *dowel, pp*, ¼ of an inch in diameter, and projecting ¾ of an inch on each side of the piece, *acb*. This pin hides another, supposed to be behind it.

3*d*. The pieces, *d* and *d''*, are each 2½ inches by 1 inch, and are shown as if just drawn off from the *dowels*, but in their true direction, i.e. at right angles to *acb*.

4*th*. The inner end of *d* is shown at *d'*, showing the two holes, 1½ inches apart, into which the *dowels* fit.

Execution.—The end view is lined as usual, leaving the dowel holes blank.

117. Ex. 2. **A dovetailed Mortise and Tenon.** Pl. VI., Fig. 58. *Mechanical Construction.*—This figure shows a species of joining called *dovetailing.* Here the mortise increases in width as it becomes deeper, so that pieces which are dovetailed together, either at right angles or endwise, cannot be pulled directly apart. The corners of drawers, for instance, are usually dovetailed; and sometimes even stone structures, as lighthouses, which are exposed to furious storms, have their parts dovetailed together.

118. *Graphical Construction.*—The sketches of this framing are arranged as two elevations. A bears the dovetail, B shows the length and breadth of the mortise, and B'' its depth. A and B belong to the same elevation.

Execution.—In this case a method is given, of representing a hidden cut surface, viz. by dotted shade lines, as seen in the hidden faces of the mortise in B''.

119. Leaving now the examples of pieces framed together at right angles let us consider :—

§ V.—*Pairs of Timbers which are framed together obliquely to each other.*

EXAMPLE 1. **A Chord and Principal.** (From a Model.) Pl. VII., Fig. 59. *Mechanical Construction.*—The oblique piece ("principal") is, as the two elevations together show, of equal width with the horizontal piece ("chord," or "tie beam"), and is framed into it so as to prevent sliding sidewise or lengthwise.

Neither can it be lifted out, on account of the bolt which is made to pass perpendicularly to the joint, *ac*, and is "chipped up" at *pp*, so as to give a flat bearing, parallel to *ac*, for the nut and bolt-head.

120. *Graphical Construction.*—1*st*. Draw *pde ;* 2*d*. Lay off *de* $=$13 inches ; 3*d*. Make *e'ea*$=30°$; 4*th*. At any point, *e'*, draw a perpendicular to *ee'*, and lay off upon it 9 inches—the perpendicular width of *e'ea ;* 5*th*. Make *ec*$=4$ inches and perpendicular to *e'e ;* bisect it and complete the outlines of the tenons, and the shoulder *aec' ;* 6*th*. To draw the nut accurately, proceed as in Pl. VI., Fig. 46–47, placing the centre of the auxiliary projection of the nut in the axis of the bolt produced, &c. (46) (96).

Execution.—*b* represents the bolt hole, the bolt being shown only on one elevation.

121. **Ex. 2. A Brace, as seen in the angle between a "post" and "girth."** (From a Model.) Pl. VII., Fig. 60. *Mechanical Construction.*—PP' is the post, GG' is the girth, and B'B" is the brace, having a truncated tenon at each end, which rests in a mortise. When the brace is quite small, it has a shoulder on one side only of the tenon, as if B'B" were sawed lengthwise on a line, *oo'*.

122. *Graphical Construction.*—To show a tenon of the brace clearly, the girth and brace together are represented as being drawn out of the post. 1*st*. Draw the post. 2*d*. Half an inch below the top of the post, draw the girth 2½ inches deep. 3*d*. From *a*, lay off *ab* and *a* each 4 inches, and draw the brace 1 inch wide. 4*th*. Make *cd* equal to the adjacent mortise ; viz. 1½ inches ; make *de*$=1$ inch, and erect the perpendicular at *e* till it meets *bc*, &c. The dotted projecting lines show the construction of B" and of the plan. At *e"* is the vertical end of the tenon *e*. On each side of *e"*, are the vertical surfaces, shown also at *cd*.

123. **Ex. 3. A Brace, with shoulders mortised into the post.** Pl. VII., Fig. 51. This is the strongest way of framing a brace. For the rest, the figure explains itself. Observe, however, that while in Fig. 59, the head of the tenon and shoulder is perpendi-

cular to the oblique piece, here, where that piece is framed into a vertical post, its head *nu* is perpendicular to the axis of the post. In Fig. 61, moreover, the auxiliary plane on which the brace alone is projected, is parallel to the length of the brace, as is shown by the situation of B, the auxiliary projection of the brace, and by the direction of the projecting lines, as *nn″*. P′ is the elevation of P, as seen in the direction *nn′*, and with the brace removed.

124. Ex. 4. **A "Shoar."** Pl. VII., Fig. 62. *Mechanical Construction.*—A " shoar " is a large timber used to prop up earth or buildings, by being framed obliquely into ·a horizontal beam and a stout vertical post. It is usually of temporary use, during the construction of permanent works ; and as respects its action, it resists compression in the direction of its length. To give a large bearing surface without cutting too far into the vertical timber, it often has two shoulders. The surface at *ab* is made vertical, for then the fibres of the post are unbroken except at *cb*, while if the upper shoulder were shaped as at *adb*, the fibres of the triangular portion *dbc* would be short, and less able to resist a longitudinal force.

125. The *Graphical construction* is evident from the figure, which is in two elevations, the left hand one showing the post only.

Execution.—The vertical surface at *ab* may, in the left hand elevation, be left blank, or shaded with vertical lines as in Pl. VI., Fig. 55*a*.

§ VI.—*Combinations of Timbers whose axes make angles of* 180° *with each other.*

126. Timbers thus framed are, in general, said to be spliced. Six forms of splicing are shown in the following figures.

EXAMPLE 1. **A Halved Splicing, pinned.** Pl. VII., Fig. 63. The *mechanical construction* is evident from the figure. When boards are lapped on their edges in this way, as in figure 69, they are said to be " *rabbetted.*"

127. *Graphical Construction.*—After drawing the lines, 12 inches apart, which represent the edges of the timber, drop a perpendicular of 6 inches in length from any point as *a*. From its lower extremity, draw a horizontal line 12 inches in length, and from *c*, drop the perpendicular *cb*, which completes the elevation. In the plan, the joint at *a* will be seen as a full line *a′a″*, and that at *b*, being hidden, is represented as a dotted line, at *b″*

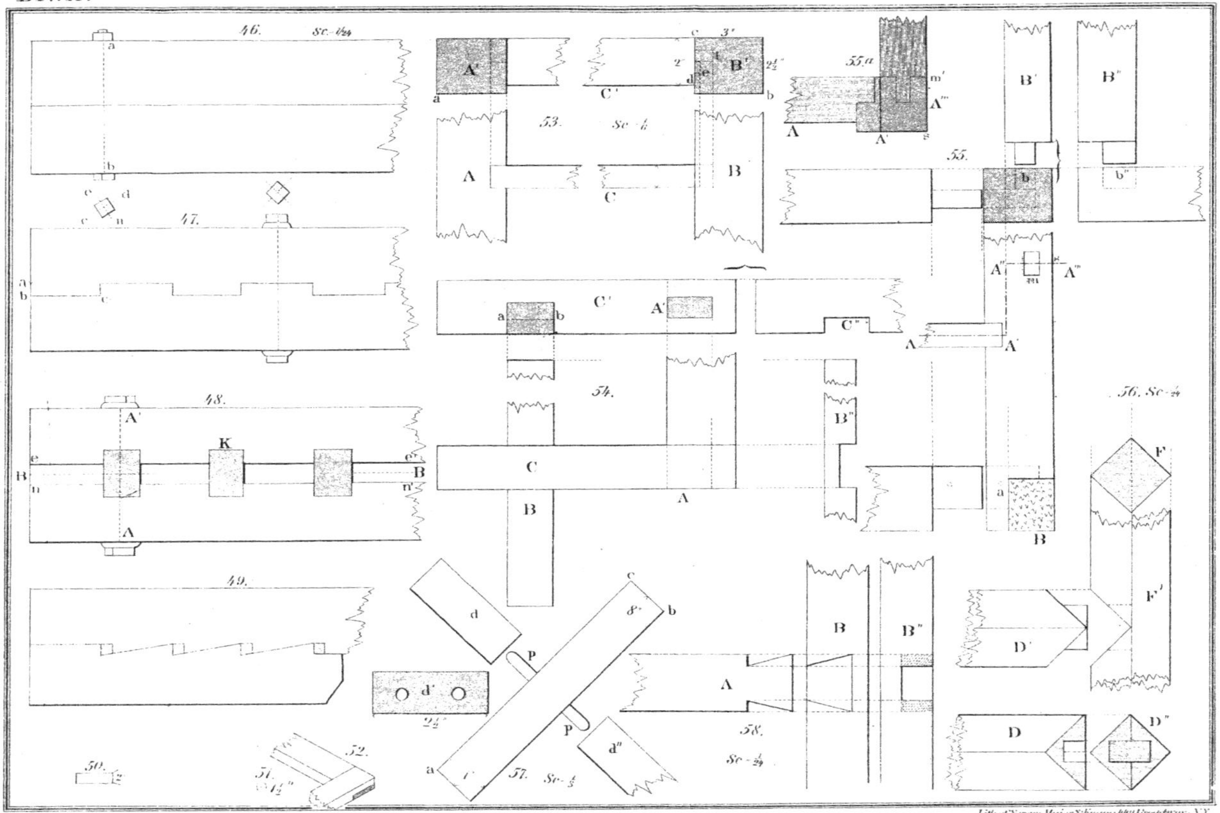
46. Sc. 1/24
47.
48.
49.
50.
51.
52.
53. Sc. 1/6
54.
55.
55 a
56. Sc. 1/24
57. Sc. 1/3
58. Sc. 1/24

Draw a diagonal, $a'b''$, divide it into four equal parts, and take the first and third points of division for the centres of two pins, having each a radius of three-fourths of an inch.

Execution.—The position of the heavy lines on these figures is too obvious to need remark.

128. Ex. 2. **Tonguing and Grooving; and Mortise and Tenon Splicing.** Pl. VII., Fig. 64. Boards united at their edges in this way, as shown in Pl. VII., Fig. 70, are said to be tongued and grooved.

Drawing, as before, the plan and elevation of a beam, a foot square, divide its depth an, at any point a, into five equal parts. Take the second and fourth of these parts as the width of the tenons, which are each a foot long.

The joint at a is visible in the plan, the one at b is not. Let $a'd$ be a diagonal line of the square $a'd$. Divide $a'd$ into three equal parts, and take the points of division as centres of inch bolts, with heads and nuts 2 inches square, and washers of $1\frac{3}{4}$ inches radius. To place the nut in any position on its axis, draw any two lines at right angles to each other, through each of the bolt centres, and on each, lay off 1 inch from those centres, and describe the nut. Project up those angles of the nut which are seen; viz. the foremost ones, make it 1 inch thick in elevation, and its washer $\frac{1}{2}$ an inch thick.

In this, and all similar cases, the head of the bolt, s, would not have its longer edges necessarily parallel to those of the nut. To give the bolt head any position on its axis, describe it in an auxiliary plan just below it.

129. Ex. 3. **A Scarfed Splicing, strapped.** Pl. VII., Fig. 65. *Mechanical Construction.*—While timbers, framed together as in the two preceding examples, can be directly slid apart when their connecting bolts are removed; the timbers, framed as in the present example, cannot be thus separated longitudinally, on account of the dovetailed form of the splice. Strapping makes a firm connexion, but consumes a great deal of the uniting material.

130. *Graphical Construction.*—After drawing the outlines of the side elevation, make the perpendiculars at a and a', each 8 inches long, and make them 18 inches apart. One inch from a, make the strap, ss', $2\frac{1}{2}$ inches wide, and projecting half an inch—i. e. its thickness—over the edges of the timber. The ear through which the bolt passes to bind the strap round the timber, projects two inches above the strap. Take the centre of the ear as the centre of the bolt, and on this centre describe the bolt head $1\frac{1}{2}$ inches square.

In the plan, the ears of the strap are at any indefinite distance apart, depending on the tightness of the nut, *s*.

A fragment of a similar strap at the other end of the scarf, is shown, with its visible ends on the bottom, near a', and back, at r, of the beam.

Execution.—The scarf is dotted where it disappears behind the strap; and so are the hidden joint at a', and the fragment of the second strap, as shown in plan. These examples may advantageously be drawn by the student on a scale of $\frac{1}{12}$. Care must then be exercised in making the large broken ends neatly, in large splinters, edged with fine ones.

131. **Ex. 4. A Scarfed Splicing; notched, keyed, and bolted.** Pl. VII., Fig. 66. *Mechanical Construction.*—This is a specimen of a very secure framing. The notches e and e' prevent lateral displacement, and the keys of hard wood give a better bearing surface than when the ends of the grain bear upon each other (99).

132. *Graphical Construction.*—Making the outlines of the timber as before, take $a'b = 3$ feet; draw the perpendicular bb', and connect a' and b'. From a' lay off 4 inches, and divide the remainder of $a'b'$ into three equal parts; then at b', and c, draw perpendiculars, each 2 inches long, above $a'b'$. Now divide $a'c$ into two equal parts, and from c and d lay off two inches towards a', and, on these distances, complete the squares which represent the keys at those points, the one at d being below $a'b'$. Lastly, draw, at a', the joint line perpendicular to $a'b'$.

In the plan, let the angles, as $eao = 60°$, a being projected down from a', and e being in the middle of the width of the plan.

Let the bolts, as nn', be so placed that their axes shall bisect $a'd$ and $b'c$; and let the washers be 5 inches in diameter, and the nuts $2\frac{1}{4}$ inches square, by $1\frac{1}{4}$ inches thick.

a', being horizontally projected at a and a'', $a''v$ is the horizontal projection of $a'v'$. Project e at e'' and draw $e''s'$, parallel to $a'v'$ till it meets $v's'$, parallel to $a'd$. Then project s' at s, and draw sv, the horizontal projection of $s'v'$. Similarly sv'' is drawn. se is the horizontal projection of $e''s$.

$a''e$ and vs will, of course, not be parallel, though this is indistinctly shown in so small a figure.

Execution.—The keys are shaded. The hidden cut surfaces of the notches are shaded in dotted shade lines, and the hidden joints are dotted.

133. **Ex. 5. A Compound Beam, with one of the component beams "fished."** Pl. VII., Fig. 67. *Mechanical Construc-*

tion.—The mode of union called *fishing*, consists in uniting two pieces, end to end, by laying a notched piece over the joint and bolting it through the longer pieces.

The figure shows this mode as applied to a compound beam, i. e. to a beam "*built*" of several pieces bolted and keyed together. The order of construction is as follows—taking for a scale $\frac{5}{8}$ of an inch to a foot.

134. *Graphical Construction.*—1*st*. The outside lines of the plan are two feet apart, the outside pieces are each $4\frac{1}{2}$ inches wide, and the interior ones $5\frac{1}{2}$ inches ; leaving four inches for the sum of the three equal spaces between the four beams.

2*d*. Let there be a joint at a'. Lay off 3 feet, each side of a' for the length of the "*fish.*"

3*d*. The straight side of this piece is let into the whole piece, b, two thirds of an inch, and into a', 1 inch, making its thickness 3 inches.

4*th*. At a', it is two inches thick, i. e. at a', the timber, a', is of its full thickness. The fish, c, is 2 inches thick for the space of one foot at each side of a'. The notches at d and e are each 1 inch deep, dd' and ee' each are one foot, and the notches at d' and e' are each 1 inch deep. The remaining portions of the fish are 1 foot long, and 3 inches wide.

5*th*. Opposite to these extreme portions, are keys, 1 foot by 3 inches, in the spaces between the other timbers, and setting an equal depth into each timber.

6*th*. In elevation, only the timber a' is seen—1 foot deep. Four bolts pass through the keys. $b'b''=5$ feet, and b' is three inches from the top, and from kk', the left hand end of the fish. $n'n''=5$ feet, and n' is 3 inches above the bottom of the timber, and 9 inches from kk'. The circular bolt head is one inch in diameter, and its washer $3\frac{1}{2}$ inches diameter and $\frac{1}{2}$ an inch thick. The thickness of the bolt head, as seen in plan, is $\frac{3}{4}$ inch. The nuts, nn, are $1\frac{1}{4}$ inches square, and $\frac{1}{2}$ an inch thick, and the bolts are half an inch in diameter.

135. The several nuts would naturally be found in various positions on their axes. To construct them thus with accuracy, as seen in the plan, one auxiliary elevation, as N, is sufficient. N, and its centre, may be projected upon as many planes—xy—as there are different positions to be represented in the plan, each plane being supposed to be situated, in reference to N, as some nut in the plan is, to its elevation. Then transfer the points on xy, &c., to the outside of the several *nut*-washers, placing the projection of the centre lines of the bolts in the plan, as lines of reference.

Execution.—The figure explains itself in this respect.

136. Ex. 6. **A vertical Splice.** Pl. VIII., Fig. 68. *Mechanical Construction.*—This splice is formed of two prongs at opposite corners of each piece, embraced by corresponding notches in the other piece. Thus in the piece B′, the visible prong, as seen in elevation, is a truncated triangular pyramid whose horizontal base is abc—$a'c'$, and whose oblique base is enc—$e'n''c''$. Besides the four prongs, two on each timber, there is a flat surface $abfq$—$c'a'q'$, well adapted to receive a vertical pressure, since it is equal upon, and common to, both timbers.

137. *Graphical Construction.*—To aid in understanding this combination, an oblique projection is given on a diagonal plane, parallel to PQ.

1*st.* Make the plan, $acfq$, with the angles of the interior square in the middle of the sides of the outer one. 2*d.* Make the distances, as $ce=2$ inches and draw en, &c. 3*d.* Make $c'n'=c'n''$ 15 inches, and draw short horizontal lines, $n''e'$, &c., on which project e, &c., after which the rest is readily completed.

138. *Execution.*—Observe, in the plans, to change the direction of the shade lines at every change in the position of the surface of the wood.

CHAPTER III.

139. Example 1. An end view of a Railroad Rail. Pl. VII., Fig. 71. *Graphical Construction.*—1*st*. Draw a vertical centre line AA′, and make AA′=3¾ inches. 2*d*. Make A′b=A′b'=2 inches. 3*d*. Make Ac=Ac'=1 inch. 4*th*. Describe two quadrants, of which $c'd$ is one, with a radius of half an inch. 5*th*. With p, half an inch from AA′, as a centre, and pd as a radius, describe an arc, dr, till it meets a vertical line through e. 6*th*. Draw the tangent rs. 7*th*. Draw a vertical line, as pt, ½ an inch from AA′, on each side of AA′. 8*th*. Bisect the angle rst' and note s', where the bisecting line meets the radius, pr, produced. 9*th*. With s' as a centre, draw the arc rt. 10*th*. At b and b' erect perpendiculars, each one fourth of an inch high. 11*th*. Draw quadrants, as $q't$, tangent to these perpendiculars and of one fourth of an inch radius. 12*th*. Draw the horizontal line tv. 13*th*. Make $nt'=nv$ and describe the arc $t'v$. 14*th*. Repeat these operations on the other side of the centre line, AA′.

Execution.—Let the construction be fully shown on one side of the centre line.

140. Ex. 2. An end elevation of a Compound Rail. Pl. VII., Fig. 72. *Mechanical Construction.*—The compound rail, is a rail formed in two parts, which are placed side by side so as to break joints, and then riveted together. As one half of the rail is whole, at the points where a joint occurs on the other half, the noise and jar, observable in riding on tracks built in the ordinary manner, are both obviated; also "chairs," the metal supports which receive the ends of the ordinary rails, may be dispensed with, in case of the use of the compound rail.

In laying a compound rail on a curve, the holes, through which the bolts pass, may be drawn past one another by the bending of the rails. To allow for this, these holes are "slotted," as it is termed, i. e. made longer in the direction of the length of the rail.

141. *Graphical Construction.*—1*st*. Make ty=4 inches. 2*d*. Bisect ty at u, and erect a perpendicular, ua, of 3½ inches. 3*d*.

Make, successively, $ur=\frac{1}{2}$ an inch; from r to $cb=2$ inches; from u to $nh=\frac{1}{8}$ of an inch; and to $ge=2\frac{3}{8}$ inches. *4th.* For the several widths of the interior parts, make $bc=\frac{3}{8}$ of an inch, and g and e each $\frac{3}{8}$ of an inch, from ua; $nh=ge$, and $ro=\frac{3}{4}$ of an inch. *5th.* To locate the outlines of the rail, make ms, the flat top, called the tread of the rail, $=2$ inches, half an inch below this, make the width, fd, 3 inches; and make the part through which the rivet passes, $1\frac{3}{4}$ inches thick, and rounded. into the lower flange which is $\frac{3}{4}$ of an inch thick.

The rivet has its axis $1\frac{5}{8}$ inches from ty. Its original head, q, is conical, with bases of—say 1 inch, and $\frac{3}{4}$ of an inch, diameter; and is half an inch thick. The other head, p, is made at pleasure, being roughly hammered down while the rivet is hot, during the process of track-laying. A thin washer is shown under this head.

142. Ex. 3. **A "Cage Valve," from a Locomotive Pump.** Pl. VII., Fig. 73–74.—*Mechanical Construction.*—This valve is made in three pieces, viz. the valve proper, Fig. 74; the cage containing it, Abb'; and the flange $bb'c$; whose cylindrical aperture—shown in dotted lines—being smaller than the valve, confines it. The valve is a cup, solid at the bottom, and makes a water tight joint with the upper surface of the flange, inside of the cage. The whole is inclosed in a chamber communicating with the pump barrel, and with the tender, or the boiler, according as we suppose it to be the inlet or outlet valve of the pump. This chamber makes a water tight joint with the circumference of the flange $bb.'$

Suppose the valve to be the latter of the two just named. The "plunger" of the pump being forced in, the water shuts the inlet valve, and raises the outlet valve, and escapes between the bars of the cage into the chamber, and from that, by a pipe, into the boiler.

143. *Graphical Construction.*—Construct the plan first, in which the bars appear as six equal and equidistant wings, which are then projected into the elevation. The diameters of the circles seen in the plan, are, in order, from the centre, $1\frac{1}{4}$, $2\frac{1}{8}$, $2\frac{1}{8}$, and 4 inches. The thickness of the cage is $\frac{3}{16}$ of an inch, and its outside height $1\frac{5}{8}$ inches, and the outside diameter $2\frac{3}{16}$ inches. The diameter of the aperture in the flange is $1\frac{3}{4}$ inches, its length $\frac{3}{4}$ of an inch, and the height of the whole $3\frac{3}{16}$ inches.

144. *Execution.*—Observe carefully the position of the heavy lines on the elevation, which conform to the rational rule of being the projections of those edges of the object itself, which divide its illuminated from its dark surfaces. The section, Fig. 74, being of metal, is appropriately shaded more finely than sections of wood.

145. Ex. 4. An oblique elevation of a Bolt Head. Pl. VII., Fig. 75. Let PQ be the intersection of two vertical planes, at right angles with each other; and let RS be the intersection, with the vertical plane of the paper to the left of PQ, of a plane which, in space, is parallel to the square top of the bolt head. On such a plane, a plan view of the bolt head may be made, showing two of its dimensions in their real size; and on the plane above RS, the thickness of the bolt head, and diameter of the bolt, are shown in their real size.

Below RS, construct the plan of the bolt head, with its sides making any angle with the ground line RS. Project its corners in perpendiculars to RS, giving the left hand elevation, whose thickness is assumed.

146. The fact that the projecting lines of a point, form, in the drawings, a perpendicular to the ground line, is but a special case of a more general truth, which may be thus stated.—When an object in space is projected upon any two planes which are at right angles to each other, the projecting lines of any point of that object form a line, in the drawing, perpendicular to the intersection of the two planes.

147. To apply the foregoing principle to the present problem; it appears that each point, as a'', of the right hand elevation, will be in a line, $a'a''$, perpendicular to PQ, the intersection of the two vertical planes of projection.

Remembering that PQ is the intersection of a vertical plane—perpendicular to the plane of the paper—with the vertical plane of the paper, and observing that the figure represents this plane as being revolved around PQ towards the left, and into the plane of the paper, and observing the arrow, which indicates the direction in which the bolt head is viewed, it appears that the revolved vertical plane, has been transferred from a position at the left of the plan *acne*, to the position, PQ, and that the centre line tu'', must appear as far from PQ as it is in front of the plane of the paper—i. e. $e''u''=eu$, showing also, that as e—e' is in the plane of the paper, its projection at e'' must be in PQ, the intersection of the two vertical planes.

Similarly, the other corners of the nut, as c'', n'', &c., are laid off either from the centre line tu'', or from PQ. Thus $v''c''=vc$, or $v'''c''=sc$. The diameter of the bolt is equal in both elevations.

148. Other supposed positions of the auxiliary plane PQ may be assumed by the student, and the corresponding construction worked out. Thus, the primitive position of PQ may be at the right of the bolt head, and that may be viewed in the opposite direction from that indicated by the arrow.

149.—Ex. 5. **A " Step " for the support of an oblique timber.** Pl. VII., Fig. 76. *Mechanical Construction.*—It will be frequently observed, in the framings of bridges, that there are certain timbers whose edges have an oblique direction in a vertical plane, while at their ends they abut against horizontal timbers, not directly, for that would cause them to be cut off obliquely, but through the medium of a prismatic block of wood or iron, so shaped that one of its faces, as ab—$n'b'$, Fig. 76, rests on the horizontal timber, while another, as ac—$e''d''c''$, is perpendicular to the oblique timber.

To secure lightness with strength, the step is hollow underneath, and strengthened by ribs, rr. The holes, $h'h''$, allow the passage of iron rods, used in binding together the parts of the bridge. These holes are here prolonged, as at h, forming tubes, which extend partly or wholly through the horizontal timber on which the step rests, in order to hold the step steadily in its place.

150. *Remark.* When the oblique timber, as T, Fig. 76 (a), sets *into*, rather than *upon*, its iron support S, so that the dotted lines, ab and cb, represent the ends of the timber, the support, S, is called a shoe.

151. *Graphical Construction.*—In the plate, abc is the elevation, and according to the usual arrangement would be placed above the plan, $e''n''c''$, of the top of the step. $a'b'e$ is the plan of the under side of the step, showing the ribs, &c. A line through nm is a centre line for this plan and for the elevation. A line through the middle point, r, of mn, is another centre line for the plan of the bottom of the step. Having chosen a scale, the position of the centre lines, and the arrangement of the figures, the details of the construction may be left to the student.

152. Ex. 6. **A metallic steam tight " Packing," for the " stuffing boxes " of piston rods.** Pl. VII., Fig. 77.

Mechanical Construction.—Attached to that end, b, of Fig. 77 (a), of a steam cylinder, for instance, at which the piston rod, p, enters it, is a cylindrical projection or " neck," n, having at its outer end a flange, ff, through which two or more bolts pass. At its inner end, at p, this neck fits the piston rod quite close for a short space. The internal diameter of the remaining portion of the neck is sufficient to receive a ring, rr, which fits the piston rod, and has on its outer edge a flange, t, by which it is fastened to the flange, ff, on the neck of the cylinder by screw bolts. The remaining hollow

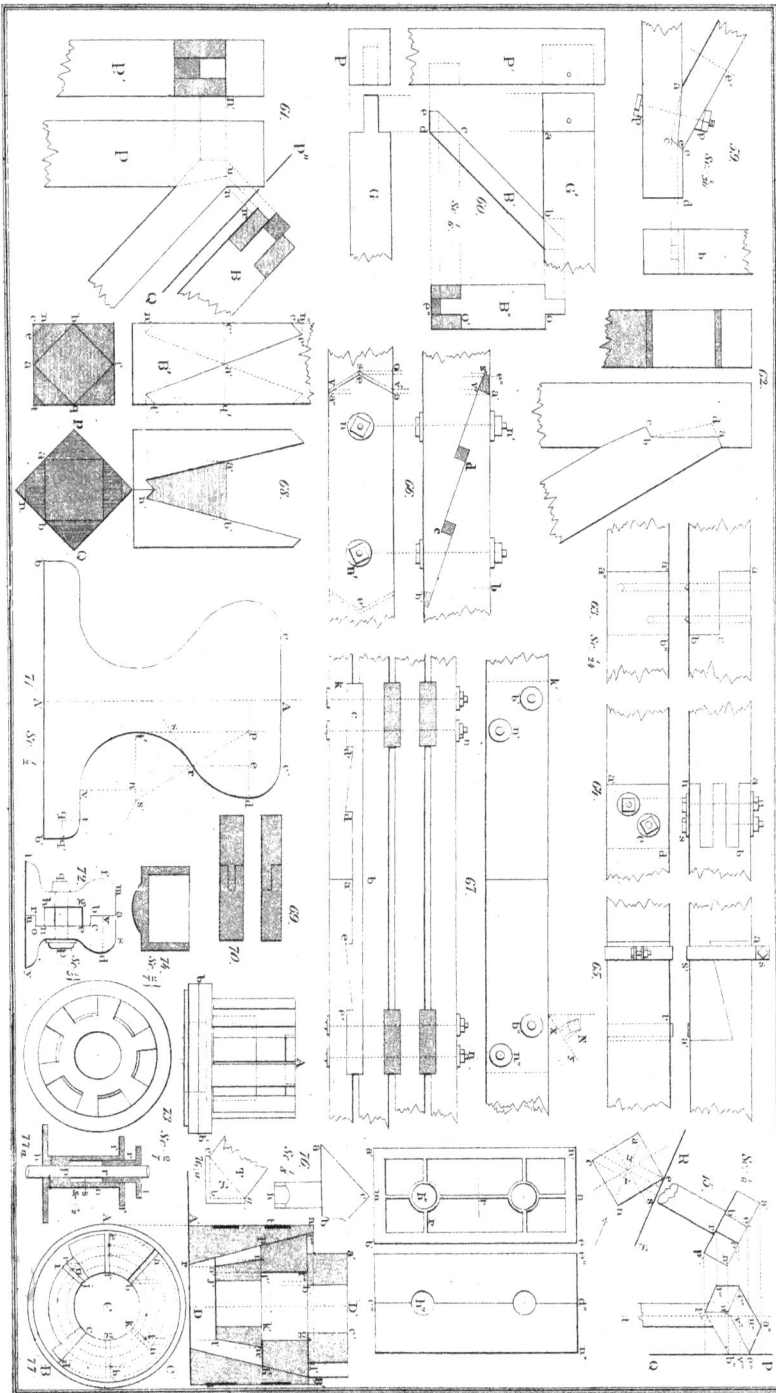

space, *s*, between the ring or " gland," *tr*, and the inner end of the
neck, is usually filled with some elastic substance, as picked hemp,
which, as held in place by the gland, *tr*, makes a steam tight joint;
which, altogether, is called the " stuffing box."

153. The objection to this kind of packing is, that it requires so
frequent renewals, that much time is consumed, for instance in rail.
road repair shops, in the preparation and adjustment of the packing.
To obviate this loss of time, and perhaps because it seems more
neat and trim to have all parts of an engine metallic, this metallic
packing, Fig. 77, was invented. ABC—A'B' is a cast iron ring,
cylindrical on the outside, and having inlaid, in its circumference,
bands, *tt'*, of soft metal, so that it may be squeezed perfectly tight
into the neck of the cylinder. The inner surface of this ring is conical, and contains the packing of block tin. This packing, as a whole,
is also a ring, whose exterior is conical, and fits the inner side of the
iron ring, and whose interior, *fkc*, is cylindrical, fitting the piston
rod closely.

154. For adjustment, this tin packing is cut horizontally into
three rings, and each partial ring is then cut vertically, as shown in
the figure, into two equal segments. *abcd—a'b'c'd'*, is one segment
of the upper ring; *efgh—e'f'f''g'h'*, is one segment of the middle ring,
and *ijkl—i'j'k'n'l'*, is one segment of the lower ring. The segments
of each ring, it therefore appears, break joints with each of the other
rings. Three of the segments, one in each partial ring, are loose,
while the other three are dowelled by small iron pins, parallel to
the axis of the whole packing.

155. *Operation.*—Suppose the interior, *fkc*, of the packing to be
of less diameter than the piston rod, which it is to surround. By
drawing it partly out of its conical iron case, the segments forming
each ring can be slightly separated, making spaces at *ab*, &c., which
will increase the internal diameter, so as to receive the piston rod.
When in this position, let the gland, *tr*, be brought to bear on the
packing, and it will be firmly held in place; then, as the packing
gradually wears away, the gland, by being pressed further into the
neck, will press the packing further into its conical seat, which
will close up the segments round the piston rod.

156. *Graphical Construction.*—Let the drawing be on a scale
of one half the original size of the packing for a locomotive valve
chest. C is the centre for the various circles of the plan, and DD',
projected up from C, is a centre line for portions of the elevation.
$Cf = \frac{9}{16}$ of an inch; $Ce = 1\frac{4}{16}$ of an inch, and $CA = 1$ inch. $A'n = 2$
inches, and $A'r = \frac{9}{16}$ of an inch. The iron case being constructed

4

from these measurements, the rings must be located so that $f''p'$, for instance, shall be $= \frac{7}{16}$ of an inch; and then let the thickness, $f'f''$, of each segment be $\frac{11}{16}$ of an inch. These dimensions and the consequent arrangements of the rings will give spaces between the segments, as at ab, of $\frac{1}{8}$ of an inch; though in fact, as this space is variable, there is no necessity for a precise measurement for it. In the plan, there are shown one segment, and a fragment, $abef$, of another, in the upper ring; one segment and two fragments of the middle ring, and both segments of the lower ring, with the whole of the iron case.

157. *Execution.*—The section lines in the elevation indicate clearly the situation of the three segments, $abcd$, $efgh$, and $ijkl$, there shown. The dark bands on the case at t and t', indicate the inlaid bands of soft metal already described.

DIVISION THIRD.

CHAPTER I.

SHADOWS.

§ I.—*Observed Facts and Practical Rules.*

158. THE lines of light, called rays, which emanate from a luminous body, proceed in straight lines. This is proved by the fact that the shadow of a straight line, as the corner of a house, is a straight line, when cast upon a plane having any position relative to the line.

159. If the student can see from his window a house, part of whose body projects beyond the rest, and a chimney upon a flat roof, he can learn, by simple inspection, many of those natural facts respecting shadows which follow directly from the preceding article, and are at the foundation of the solution of most of the simple problems which occur in roof or bridge drawing, where the longer shadows are most frequently cast by *straight lines* on *plane surfaces.*

160. It will be thus observed that the shadow of a vertical edge, ab, Pl. VIII., Fig. 78, of the body of the house will be a vertical line, $a'b'$, on the front wall of the wing behind it; that the shadow of a horizontal line, as be—the arm for a swinging sign—which is parallel to the wing wall, will be a horizontal line, $b'e'$, parallel to be; that the shadow of a horizontal line, bc, which is perpendicular to the wing wall, will have an oblique shadow, cb', commencing at c, where the line pierces the wing wall, and ending at b', where a ray of light through b pierces the wing wall; and finally that the shadow of a *point*, b, is at b' where the ray bb', through that point, pierces the surface receiving the shadow.

161. Passing now to Pl. VIII., Fig. 79, which represents a chim-

ney upon a flat roof, we observe that the shadows of bc and cd—lines parallel to the roof—are $b'c'$ and $c'd'$, lines equal to, and parallel to, the lines bc and cd; and that the shadows of ab and ed are ab' and ed'—similar to the shadow cb' in Fig. 78—i. e. commencing at a and e, where the lines casting them meet the roof, and ending at b' and d', where *rays* through b and d meet the roof.

162. By considering the facts expressed in the two preceding articles, we may arrive at three simple general principles; viz. 1*st*. The shadow of a point on any surface, is where a ray of light drawn through that point meets that surface. 2*nd*. The shadow of a straight line on a plane, parallel to it, is a parallel straight line; and 3*rd*. The shadow of a line which is perpendicular to a plane, is in the *direction of the projection* of a ray of light on that plane; for in Fig. 78, bc, being perpendicular to the wing wall, is a projecting line of the ray bb' upon that wall, and b', where the ray meets the wall, is another point of its projection upon the wall; hence cb' is the projection of the ray bb' on the wing wall. So in Pl. VIII., Fig. 79, the planes of the triangles abb' and edd' are perpendicular to the top of the roof; hence the shadow lines ab' and ed' are projections of the rays bb' and dd', which limit those shadows.

163. By further examination, the following truths become apparent; 1*st*. The shadow of a circle on a plane parallel to itself, is an equal circle. 2*nd*. The point, as at a', Pl. VIII., Fig. 78, where a shadow—aa'— on one surface intersects a second surface, is a point of the shadow of the same line—ab, on the second surface.

164. Passing now to the graphical solution of problems of shadows, there appear certain necessary conditions to be known, and certain practical rules to be observed.

The necessary conditions for solving a problem in shadows are three:—1*st*, the position of the body casting the shadow; 2*nd*, the position of the surface receiving the shadow; and 3*rd*, a given direction of the light.

165. In respect to the above conditions, it may be remarked:— 1*st*, that the body casting the shadow is given by its projections, made according to the methods of DIVISION I.; 2*nd*, that the surfaces, one or more, which receive the shadow may be the planes of projection themselves, or other surfaces represented by their projections on those planes; and 3*rd*, that the usual assumed direction of the light is such, that *its projections* make angles of 45° with the ground line.

166. The direction of the light itself, corresponding to the direc-

tion of its projections just mentioned, may be understood from an inspection of Pl. VIII., Fig. 80.

Let a cube be placed so that one of its faces, $L'L_1b$, shall coincide with the vertical plane, and another face, $L''aL_1$, with the horizontal plane. The diagonals, $L'L_1$ and $L''L_1$, of these faces will be the *projections* of a ray of light, and the diagonal, LL_1, of the cube will be the ray itself; for the point of which L' and L'' are the projections must be in each of the projecting perpendiculars $L'L$ and $L''L$; hence at L, their intersection.

167. Referring now more particularly to the practical rules above mentioned (Art. 164), it appears—*1st*, that as two points determine a straight line, we can determine the shadow of a straight line by passing rays through any two of its points, and finding where those rays pierce the surface receiving the shadow; *2nd*, that as one point will determine a line whose direction is already known, we have only to operate as above with a single ray, when the line casting the shadow is parallel, or perpendicular, to the plane receiving the shadow.

§ II.—*Problems.*

168. PROB. 1. *To find the shadow of a vertical beam, upon a vertical wall.* Pl. VIII., Fig. 81. Let AA' be the beam, let the vertical plane of projection be taken as the vertical wall, and let the light be indicated by the lines, as ab—$a''b'$. The edges which cast the visible shadow are a—$a'a''$; ac—a''; and ce—$a''e'$. The shadow of a—$a'a''$ is a vertical line from the point b', which point is where the ray from a—a'' pierces the vertical plane. ab—$a''b'$ pierces the vertical plane in a point whose horizontal projection is b. b' must be in a perpendicular to the ground line from b (Art. 15), and also in the vertical projection, $a''b'$, of the ray, hence at b'. $b'b$ is therefore the shadow of a—$a''a'$. The shadow of ac—a'' is the line $b'd'$, limited by d', the shadow of the point c—a'' (162). The shadow of ce—$a''e'$ begins at d', and is parallel to ce—$a''e'$, but is partly hidden.

169. *Execution.*—The boundary of a shadow being determined, its surface is, in practice, indicated by shading, either with a tint of indian ink, or by parallel shade lines. The latter method, affording useful pen practice, may be profitably adopted.

170. PROB. 2. *To find the shadow of an oblique timber, which is parallel to the vertical plane, upon a similar timber resting against the back of it.* Pl. VIII., Fig. 82. Let AA' be the timber which

casts a shadow on BB', which slants in an opposite direction. The edge ac—$a'c'$ of AA', casts a shadow, parallel to itself, on the front face of BB'; hence but one point of this shadow need be constructed. Two, however, are found, one being a check upon the other. Any point, aa', taken at pleasure in the edge ac—$a'c'$ casts a point of shadow on the front plane of BB', whose horizontal projection is b, and whose vertical projection (see Prob. 1) must be in the ray $a'b'$, and in a perpendicular to the ground line at b; hence it is at b'. The shadow of ac—$a'c'$ being parallel to that line, $b'd'$ is the line of shadow. d', the shadow of c—c', was found in a similar manner to that just described.

It makes no difference that b' is not on the actual timber, BB'; for the face of that timber is but a *limited physical plane*, forming a portion of the *indefinite immaterial plane*, in which bb' is found; hence the point b' is as good for finding the direction of the indefinite line of shadow, $b'd'$, as is d', on the timber BB', for finding the *real portion* only of the line of shadow, viz. the part which lies across BB'.

Observe that the back upper edge of AA', which is a heavy line, as seen in the plan, is so represented only till the intersection, near b—c', of the two timbers.

171. PROB. 3. *To find the shadow of a fragment of a horizontal timber, upon the horizontal top of an abutment on which the timber rests.* Pl. VIII., Fig. 83. Let AA' be the timber, and BB' the abutment. The vertical edge, a—$a''a'$, casts a shadow, ab, in the direction of the horizontal projection of a ray (162, *3rd*), and limited by the shadow of the point a—a'. The shadow of a—a' is at bb', where the ray ab—$a'b'$ pierces the top of the abutment; b' being evidently the vertical projection of this point, and b being both in a perpendicular, $b'b$, to the ground line and in ab, the horizontal projection of the ray. The shadow of ad—a' is bb'', parallel to ad—a', and limited by the ray $a'b'$—db''. The shadow of de—$a'e'$ is $b''c$, parallel to de—$a'e'$, and limited by the edge of the abutment.

172. PROB. 4. *To find the shadow of an oblique timber, upon a horizontal timber into which it is framed.* Pl. VIII., Fig. 84. The upper back edge, ca—$c'a'$, and the lower front edge through ee', of the oblique piece, are those which cast shadows. By considering the point, c, in the shadow of bc, Fig. 78, it appears that the shadow of ac—$a'c'$, Fig. 84, begins at cc', where that edge

pierces the upper surface of the timber, BB', which receives the shadow. Any other point, as aa', casts a shadow, bb', on the plane of the upper surface of BB', whose vertical projection is evidently b', the intersection of the vertical projection, $a'b'$, of the ray $ab—a'b'$ and the vertical projection, $e'c'$, of the upper surface of BB', and whose horizontal projection, b, must be in a projecting line, $b'b$, and in the horizontal projection, ab, of the ray. Likewise the line through ee', and parallel to cb, is the shadow of the lower front edge of the oblique timber upon the top of BB'.

This shadow is real, only so far as it is actually on the top surface of BB', and is visible and therefore shaded, only where not hidden by the oblique piece. Where thus hidden, its boundary is dotted, as shown at ea.

The point, bb', is in the plane of the top surface of BB', produced.

173. PROB. 5. *To find the shadow of the side wall of a flight of steps, upon the faces of the steps.* Pl. VIII., Fig. 85. The steps can be easily constructed in good proportion, without measurements, by making the height of each step two thirds of its width, taking four steps, and making the side walls rectangular parallelopipedons.

The edges, $aa''—a'$ and $a—ra'$, of the left hand side wall are those which cast shadows on the steps.

The former casts horizontal shadow lines on the tops of all the steps, from the upper one till the one containing the shadow of the point aa'; and the latter casts vertical shadow lines on the front faces of the steps, from the lower one till the one containing the shadow of aa'.

The shadow of aa' falls at dd' on the top of the step 3—3, as found by trial, i. e., a projecting line from no other point than d' of the elevation will meet the ray ad on the top of a step; and a projecting line from no point, as g, e, &c., of the plan will meet the vertical projection, $a'd'$, of this ray on the vertical face of a step. The shadow of the point aa' having been found, the other shadow lines are readily drawn as already described, and as indicated by the diagram.

Let the student, to make a special exercise, vary the proportion of the steps or the direction of the light, so as to bring the shadow of aa' upon the vertical face of a step.

174. PROB. 6. *To find the shadow of a short cylinder, or washer,*

upon the vertical face of a board. Pl. VIII., Fig. 86. Since the circular face of the washer is parallel to the vertical face of the board, BB', its shadow will be an equal circle (163), of which we have only to find the centre, OO'. This point will be the shadow of the point CC' of the washer, and is where the ray CO—C'O' pierces the board BB'. The elements, rv—r' and tu—t', of the cylindrical surface have the tangents $r'r''$ and $t't''$ for their shadows. These tangents, with the semicircle $t''r''$, make the complete outline of the required shadow.

175. PROB. 7. *To find the shadow of a nut, upon a vertical surface, the nut having any position.* Pl. VIII., Fig. 86. Let $a'c'e'$—ace be the projections of the nut, and BB' the projections of the surface receiving the shadow. The edges, $a'c'$—ac and ce—$c'e'$, of the nut cast shadows parallel to themselves, since they are parallel to the surface which receives the shadow. $a'n'$—an are the two projections of the ray which determines the point of shadow, nn'; $c'o'$—co are the projections of the ray used in finding oo', and $e'r'$—er is the ray which gives the point of shadow, rr'. The edges at aa' and ee', which are perpendicular to BB', cast shadows, $a'n'$ and $e'r'$, in the direction of the projection of a ray of light on BB'. (See cb', the shadow of cb, Pl. VIII., Fig. 78.)

176. PROB. 8. *To find the shadow of a vertical cylinder, on a vertical plane.* Pl. VIII., Fig. 87. The lines of the cylinder, CC', which cast visible shadows, are the element a—$a''a'$, to which the rays of light are tangent, and a part of the upper base. The shadow of a—$a''a'$, is gg', found by the method given in Prob. 1. At g', the curved shadow of the upper base begins. This is found by means of the shadows of several points, bd', cc', dd', &c. Each of these points of shadow is found as g was, and then they are connected by hand, or by the aid of the curved ruler.

It is well to construct one invisible point, as u', of the shadow, to assist in locating more accurately the visible portion of the curved shadow line.

177. PROB. 9. *To find the shadow of a horizontal beam, upon the slanting face of an oblique abutment.* Pl. VIII., Fig. 88. The simple facts illustrated by Pl. VIII., Figs. 78–79, have no reference to the case of surfaces of shadow, other than vertical or horizontal. But they illustrate the fact that the point where a line (see bc, Fig. 78) pierces a surface, is a point of the shadow of the line

upon such surface. To solve this problem, therefore, in a purely
elementary manner, we will proceed indirectly, i. e., by finding the
shadow on the horizontal top of the abutment and on its horizontal base. The points where these shadows meet the front
edges of this top and this base, will be points of the shadow on
the slanting face, *qne*. (See *a'*, Fig. 78.)

By Prob. 3 is found *gc*, the shadow of the upper back edge,
ax—a'x', of the timber, AA', upon the top of the abutment. *c*, the
point where it meets the front edge, *ec*, is a point of the shadow of
AA' on the inclined face. By a similar construction with any ray,
as *bp—b'p'*, is found *qp*, the shadow of *c'b'—db* upon the base of
the abutment; and *q*, where it intersects *nq*, is a point of the shadow of *db—c'b'* on the face, *qne*. The point, *dd'*, where the edge,
db—c'b', meets *dc*, is another point of the shadow of that edge;
hence *dq—d'q'* is the shadow of the front lower edge, *db—c'b'*, on
the inclined face of the abutment. The line through *cc'*, parallel to
dq—d'q', is the shadow of the upper back edge *ax—a'x'*, and completes the solution.

178. PROB. 10. *To find the shadow of a pair of horizontal timbers, which are inclined to the vertical plane, upon that plane.*
Pl. VIII., Fig. 89. Let the given bodies be situated as shown in
the diagram. In the elevation we see the thickness of one timber
only, because the two timbers are supposed to be of equal thickness
and halved together. As neither of the pieces is either parallel or
perpendicular to the vertical plane, we do not know, in advance,
the direction of their shadows. It will therefore be necessary to
find the shadows of two points of one edge, and one, of the diago-
nally opposite edge, of each timber. The edges which cast shadows
are *ac—a'c'* and *ht—e'h'*, of one timber, and *ed—e'k* and *mv—a'm"*
of the other. All this being understood, it will be enough to point
out the shadows of the required points, without describing their
construction. (See *bb'*, Fig. 81.) *bb'* is the shadow of *aa'*, and *dd'*
is the shadow of *cc';* hence the shadow line *b'—d'* is determined.
So *uu'* is the shadow of *hh';* hence the shadow line *u'w'* may be
drawn parallel to *b'd'*. Similarly, for the other timber, *ff'* is the
shadow of *ee'*, and *oo'* of *mm'*. One other point is necessary, which
the student can construct. The process might be shortened somewhat by finding the shadows of the points of intersection, *p* and *r*,
which would have been the points *p"* and *r"* of the intersection of
the shadows, and thus, points common to both shadows.

179. Prob. 11. *To find the shadow of a pair of horizontal timbers, which intersect as in the last problem, upon the inclined
face of an oblique abutment.* Pl. VIII., Fig. 90. The opening
remarks of Prob. 9 apply equally well to this present one; and the
methods of solution being the same in both, only the peculiar fea-
tures of the present problem will be noticed. Since the timber, BB′,
lies in the same direction with the light, the intersection of its two
vertical faces with the top of the abutment are its shadows on that
top, and *oo′* and *vv′*, where those shadows meet *cn—c′n′*, the front
edge of the abutment, are points of the shadow on the longer oblique
face of the abutment. The points *r′* and *t′*, on the lower edge of the
same slanting face, are similarly found, and *o′v′r′t′* is the required
shadow of BB′.

The shadow line, *en*, of the piece AA′ on the top of the abutment
is found as was *gc* in Fig. 88. *n—n′* is a point of the shadow of
AA′ on the inclined face of the abutment. It is necessary to find
two points of the shadow of its lower front edge, *bc—b′c′*. *cc′* is
one of these points, being the point where the edge *bc—b′c′* meets
the edge *cn—c′n′* of the abutment. In attempting to find a point
of shadow on the lower edge, *t′r′*, of the abutment by finding an
indefinite shadow line, like *pq*, Fig. 88, on its lower base, we meet
with the inconvenience of finding this shadow to be very far out of
the central portion of the figure, so that a point, as *q*, in Fig. 88,
will not be below the ground line.

To meet this case, we make use of one of those graphic artifices,
which are not theoretically necessary, but practically convenient in
rendering the diagram compact. Thus we take an intermediate
auxiliary horizontal plane, as *s′d′*, at about half the height of the
abutment, and find its intersection, *s′d′—sq*, with the abutment,
which, in the plan, will be projected from *s′d′* at *sq*, about half way
between the top and bottom edges of the abutment.

The shadow of the point, *bb′*, in the edge of the timber, is *dd′*, on
this auxiliary plane, and *ds* is the indefinite shadow of the lower
front edge of the timber on the same plane; then *ss′* is a point in
the required shadow on the inclined face of the abutment; and by
drawing *c′s′* and *n′p*, parallel to it, we have the required shadow in
vertical projection. The horizontal projection is omitted, to avoid
confusing the diagram. It is contained between a line joining *c* and
s, and a line through *n*, parallel to *cs*.

180. Prob. 12. *To find the shadow of the floor of a bridge
upon a vertical cylindrical abutment.* Pl. VIII., Fig. 91. The

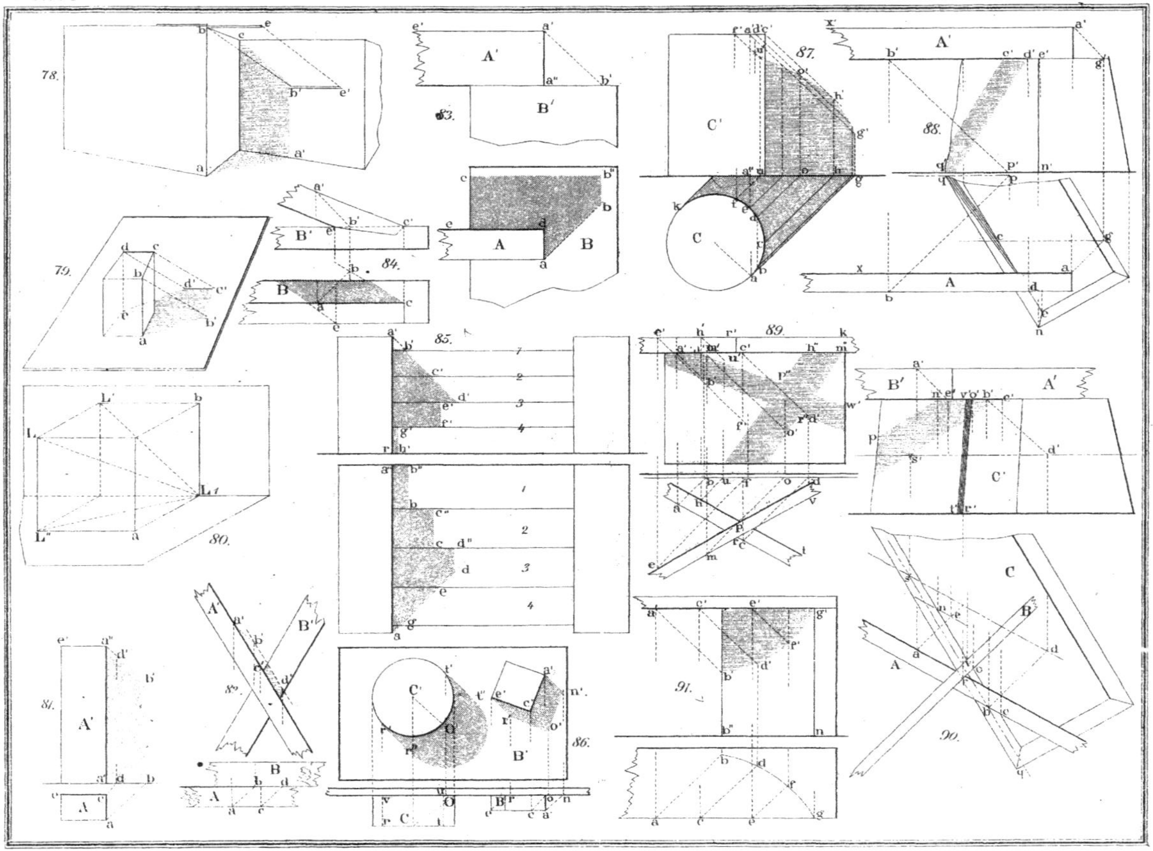

line ag—$a'g'$ is the edge of the floor which casts the shadow. bdg—$b'e'g'n$ is the concave vertical abutment receiving the shadow. gg', where the edge ag—$g'a'$ of the floor meets this curved wall, is one point of the shadow. f is the horizontal projection of the point where the ray, ef—$e'f'$, meets the abutment; its vertical projection is in $e'f'$, the vertical projection of the ray, and in a perpendicular to the ground line, through f, hence at f'. Similarly we find the points of shadow, dd' and bb', and joining them with f' and g', have the boundary of the required shadow. Observe, that to find the shadow on any particular vertical line, as b—$b''b'$, we draw the ray in the direction b—a; then project a at a', &c.

Remark.—The student may profitably exercise himself in changing the positions of the given parts, while retaining the methods of solution now given.

For example, let the parts of the last problem be placed side by side, as two elevations, giving the shadow of a vertical wall on a horizontal concave cylindrical surface; or, let the timbers, Fig. 89, be in vertical planes, and let their shadows then be found on a horizontal surface.

CHAPTER II.

181. *Notes.*—1. It is not proposed here to go largely into the theory of shading in general, but to make a special study of each one of the four practically useful exercises here given, and to notice only such principles as are involved in their elaboration.

2. The general problem for the author has been, on the one hand, to give nothing " by rote," or arbitrarily, but rather on rational grounds ; and on the other hand, to discover methods so elementary, and solutions so simple in form, as to render the problems of this DIVISION of the course perfectly intelligible to the student at this early stage of his general course of graphical study.

182. EXAMPLE 1°. *To shade the elevation of a vertical right hexagonal prism, and its shadow on the horizontal plane.* Pl. IX., Fig. 92. Let the prism be placed as represented, at some distance from the vertical plane, and with none of its vertical faces parallel to the vertical plane. The face, A, of the prism is in the light ; in fact, the light strikes it nearly perpendicularly, as may be seen by reference to the plan ; hence it should receive a very light tint of indian ink. The left hand portion of the face, A, is made slightly darker than the right hand part, it being more distant ; for the reflected rays, which reach us from the left hand portion, have to traverse a greater extent of air than those from the neighborhood of the line tt', and hence are more absorbed or retarded ; or, without attempting to say what happens to these rays, the fact is, that they make a weaker impression on the eye, causing the left hand portion of A, from which they come, to appear darker than at tt'.

183. *Remarks.*—a. It should be remembered that the whole of face A is very light, and the difference in tint between its opposite sides very slight.

b. As corollaries from the preceding, it appears : *first*, that a surface parallel to the vertical plane would receive a uniform tint throughout ; and, *second*, that of a series of such surfaces, all of which are in the light, the one nearest the eye would be lightest, and the one furthest from the eye, darkest.

c. It is only for great differences in distance that the above effects are manifest in nature; but the whole system of drawing by projec·tions being an artificial one, both in respect to the shapes which it gives in the drawings, and in the absence of surrounding objects which it allows, we are obliged to exaggerate natural appear-ances in some respects, in order to convey a clearer idea of the forms of bodies.

d. The mere manual process of shading small surfaces is here briefly described. With a sharp-pointed camel's-hair brush, wet with a very light tint of indian ink, make a narrow strip against the left hand line of A, and soften off its edge with another brush slightly wet with clear water. When all this is dry, commence at the same line, and make a similar but wider strip, and so proceed till the whole of face A is completed, when any little irregularities in the gradation of shade can be evened up with a delicate sable brush, *damp* only with very light ink.

184. Passing now to face B, it is, as a whole, a little darker than A, because, as may be seen by reference to the plan, while a beam of rays of the thickness *mp* strikes face A, a beam having only a thick-ness *pv*, strikes face B ; i. e., we assume, *first*, that the *actual bright-ness* of a flat surface is proportioned to the number of rays of light which it receives; and, *second*, that its *apparent brightness* is, other things being the same, proportioned to its actual brightness. Also, the part at *a—a'* being a little more remote than the line *t—t'*, the part at *a—a'* is made a very little darker.

185. The face C is decidedly darker than B, as it receives no light, except the small amount which it receives by reflection from sur-rounding objects. This side, C, is *darkest* at the edge *a—a'* which is nearest to the eye. This statement is confirmed by experience ; for while the shady side of a house near to us appears in strong con-trast with the illuminated side, the shady side of a remote building appears scarcely darker than the illuminated side. This fact may be accounted for as follows. The air itself is, to some extent, a reflecting medium, which is intensely illuminated by the sun; some of its reflections are cast in the direction of the eye, *and the eye refers to the distant body all the light which comes in the direction of that body.* Now, the more distant the body, the greater will be the accumulation of reflected rays added to the few straggling rays which the shady side of the body remits to the eye; hence the more distant the shady part of a body, the lighter it appears. It may be

objected that this is "special pleading," since it would make out the *remoter* parts of illuminated surfaces as the lighter parts. But not so; for the air is a nearly perfect transparent medium, and hence reflects but little, compared with what it transmits to the opaque body; but being not quite transparent, it absorbs the reflected rays from the distant body, in proportion to the distance of that body, making therefore the remoter portions darker; while the *very weak* reflections from the shady side are *reinforced* or replaced by more of the *comparatively stronger* atmospheric reflections, in case of the remote, than in case of the near part of that shady side. Thus is made out a consistent theory.

In relation, now, to the shadow, it will be lightest where furthest from the prism, since the atmospheric reflections evidently have to traverse a less depth of darkened air in the vicinity of *de—d'* than near the *lower* base of the prism at *abc*.

186. Ex. 2°. *To shade the elevation of a vertical cylinder.* Pl. IX., Fig. 93. Let the cylinder stand on the horizontal plane. The figures on the elevation suggest the comparative depth of color between the lines adjacent to the figures. The reasons for so distributing the tints will now be given.

The darkest part of the figure may properly be assumed to be that to which the rays of light are tangent; viz., the vertical line at *tt'*, from which the tint becomes lighter in both directions.

The lightest line will be at the place which reflects the most light to the eye. Now it is a principle of optics that the incident and the reflected ray make equal angles with a perpendicular to a surface. But *n*C is the incident and C*e* the reflected ray, to the eye; hence at *d* is the point, on the surface, where they make equal angles with the perpendicular (normal) *d*C, and the vertical line at *d—d'* is the lightest line.

187. *Remark.*—The question may here arise, "If all the light that is reflected towards the eye is reflected from *d*—as it appears to be—how can any other point of the body be seen?" To answer this question requires a notice—*first*, of the difference between polished and dull surfaces; and, *second*, between the case of light coming *wholly* in one direction, or *principally* in one direction. If the cylinder CC', considered as perfectly polished, were deprived of all reflected light from the air and surrounding objects, the line at *d—d'* would reflect to the eye all the light that the body would remit towards the eye, and would appear as a line of brilliant light, while other parts remitting no light whatever would be totally

invisible. Let us now suppose a reflecting medium, though an imperfect one, as the atmosphere, to be thrown around the body. By reflection, every part of the body would receive some light from all directions, and so would remit some light to the eye, making the body visible, though faintly so. But no body has a polish that is absolutely perfect; rather, the great majority of those met with in engineering art have entirely dull surfaces. Now a dull surface, greatly magnified, may be supposed to have a structure like that shown in Pl. IX., Fig. 97, in which many of the asperities may be supposed to have one little facet each, so situated as to remit to the eye a ray received by the body directly from the principal source of light.

188. Having thus proved that a cylinder placed before our eyes can be seen, we may proceed with an explanation of the distribution of tints. b is midway between d and t. At b, the ink may be diluted, and at $e—e'$, much more diluted, as the gradation from a faint tint at e to absolute whiteness at d should be without any abrupt transition anywhere.

The beam of incident rays which falls on the segment dn, is broader than that which strikes the equal segment de; hence the segment nd is, on the elevation, marked 5, as being the lightest band which is tinted at all. The segment nr, being a little more obliquely illuminated, is less bright, and in elevation is marked 3, and may be made darkest at the left hand limit. Finally, the segment rv receives about as much light as rn, but reflects it within the very narrow limits, s, hence appears brighter. This condensed beam, s, of reflected rays would make rv the lightest band on the cylinder, but for two reasons; *first*, on account of the exaggerated effect allowed to increase of distance from the eye; and, *second*, because some of the asperities, Fig. 97, would obscure some of the reflected rays from asperities still more remote; hence rv is, in elevation, marked 4, and should be darkest at its right hand limit.

189. The process of shading is the same as in the last exercise. Each stripe of the preliminary process may extend past the preceding one, a distance equal to that indicated by the short dashes at the top of Fig. 94. When the whole is finished, there should be a uniform gradation of shade from the darkest to the lightest line, free from all sudden transitions and minor irregularities.

190. Ex. 3°. *To shade a right cone standing upon the horizontal plane, together with its shadow.* Pl. IX., Fig. 95. The shadow of the cone on the horizontal plane, will evidently be bounded

by the shadows of those lines of the convex surface, at which the light is tangent. The vertex is common to both these lines, and casts a shadow, v'''. The shadows being cast by straight lines of the conic surface, are straight, and their extremities must be in the base, being cast by lines of the cone, which meet the horizontal plane in the cone's base; hence the tangents $v'''t$ and $v'''t''$ are the boundaries of the cone's shadow on the horizontal plane, and the lines joining t' and t'' with the vertex are the lines to which the rays of light are tangent; i. e., they are the darkest lines of the shading; hence tv, the visible one in elevation, is to be vertically projected at $t'v'$.

The lightest line passes from vv' to the middle point, y, between n and p in the base. At q and at p a change in the darkness of the tint is made, as indicated by the figures seen in the elevation. In the case of the cone, it will be observed that the various bands of color are triangular rather than rectangular, as in the cylinder; so that great care must be taken to avoid filling up the whole of the upper part of the elevation with a dark shade.

191. Ex. 4°. *To shade the elevation of a sphere.* Pl. IX., Fig. 96. It is evident that the parallel tangents to a sphere, at all the points of a great circle, will be a system of parallel lines which will be perpendicular to that circle. Hence, all the space around the sphere being filled with parallel rays, some ray will be found tan_ gent to the sphere at each point of a great circle, perpendicular to those rays. Such a great circle will be the curve of darkest shade on the sphere. Next, let us consider, that every circle, whether great or small, which is cut from the sphere in the same direction as that in which the light proceeds, will have two rays tangent to it on opposite sides, whose points of tangency must be points of the circle of shade; being points in which rays of light are tangent to the surface of the sphere.

Next, let us recollect that always, when a line is parallel to a plane, its projection on that plane will be seen in its true direction. Now BD' being the direction of the light, as seen in elevation, let BD' be the trace, on the vertical plane, of projection—taken through the centre of the sphere—of a new plane perpendicular to the vertical plane, and therefore parallel to the rays of light. The projection of a ray of light on this plane, BD', will be parallel to the ray itself, and therefore the angle made by this projection with the trace BD' will be equal to the angle made by the ray with the vertical plane. But, referring to Pl. VIII., Fig. 80, we see that in the triangle

LL′L₁, containing the angle LL₁L′ made by the ray LL₁ with the vertical plane, the side L′L₁ is the hypothenuse of the triangle L′b L₁, each of whose other sides is equal to LL′. Hence in Pl. IX., Fig. 96, take any distance, Bc, make AB perpendicular to BD′, and on it lay off BD = Bc, then make BD′ = Dc, join D and D′, and DD′ will be the projection of a ray upon the plane BD′, and BD′D will be the true size of the angle made by the light with the vertical plane; it being understood that the plane BD′, though in space perpendicular to the vertical plane, is, in the figure, represented as revolved over towards the right till it coincides with the vertical plane of projection, and with the paper.

192. We are now ready to find points in the curve of shade. oo′ is the vertical projection of a small circle parallel to the plane BD′, and also of its tangent rays. The circle og′o′, on oo′ as a diameter, represents the same circle revolved about oo′ as an axis and into the vertical plane of projection. Drawing a tangent to og′o′, parallel to DD′, we find g′, a visible point of the curve of shade, which, when the circle revolves back to the position oo′, appears at g, since, as the axis oo′ is *in* the vertical plane, an arc, g′g, described about that axis, must be *vertically* projected as a straight line. (See Art. 39.)

In a precisely similar manner are found h, k, m, and f. At A and B, rays are also evidently tangent to the sphere. Through A, f, &c., to B the visible portion of the curve of shade may now be sketched.

193. The most highly illuminated point is 90° distant from the great circle of shade; hence, on ABQ, the revolved position of a great circle which is perpendicular to the circle of shade, lay off k′Q = qB = the chord of 90°, and revolve this perpendicular circle back to the position qq′, when Q will be found at r′. But the brilliant point, as it appears to the eye, is not the one which *receives* most light, but the one that *reflects* most, and this point is midway, in space, between r′ and r, i. e. at P, found by bisecting QB, and drawing RP; for at R the incident ray whose revolved position is Qr, parallel to DD′, and the reflected ray whose revolved position is rB, make equal angles with R′R, the perpendicular (normal) to the surface of the sphere.

194. In regard now to the second general division of the problem —the distribution of tints; a small oval space around P should be left blank. The first stripe of dark tint reaches from A to B, along the curve of shade, and the successive stripes of the same tint extend to BqA on one side of BkA, and to oew on the other. Then take

a lighter tint on the lower half of the next zone, and a still lighter one on its upper half (2) and (3). In shading the next zone, use an intermediate tint (3-4), and in the zone next to P a very light tint on the lower side (4), and the lightest of all on the upper side (5). After laying on these preliminary tints, even up all sudden transitions and minor irregularities as in other cases.

DIVISION FOURTH.

ISOMETRICAL DRAWING.

CHAPTER I.

FIRST PRINCIPLES.

195. FROM what has been seen in the course on Projections (DIVISION FIRST), it appears that when two lines are equal, and both parallel to the planes of projection, as in Pl. I., Figs. 1, 2, their projections are equal. This has also frequently been illustrated in the problems of DIVISION SECOND.

196. It has also been seen, that while a line viewed perpendicularly, i. e., a line parallel to the planes of projection, is seen in its true size,—one which is viewed obliquely, appears shorter than it really is. This has been repeatedly shown by comparison of the two projections of the edges of a nut, in the problems of DIVISION SECOND.

197. From the foregoing, it is not difficult to see that if *any* two lines be equal and parallel to each other in space, their projections will also be equal and parallel to each other. This is easily proved, however; for as the lines in space are equal and parallel, the distance between the projecting lines of one of them is equal to the distance between the projecting lines of the other. Hence the projections of equal and parallel lines are equal and parallel—to each other—but not to the lines in space, when the latter are oblique to the planes of projection.

198. But it is not the *parallelism*, but the *equal inclination* that makes the projections *equal;* hence *equal* lines, *equally inclined* to the planes of projection, will have equal projections. Conversely, if equal projections are the projections of equal lines, those lines must be equally inclined to the planes of projection.

199. To apply the foregoing principles to the development of the elements of Isometrical Drawing, let us take a cube, and begin by supposing that we see it in right vertical projection, as in Pl. IX., Fig. 98. In this case, but one side of the cube will be visible, and that will appear as a square.

Next, suppose the cube to be turned horizontally, until two of its vertical faces should become visible. Evidently it can be turned just so far as to show those two faces equally, as in Pl. IX., Fig. 99, where a—a' and b—b' are equal.

If now, without turning the cube any more to the right or left, its upper back corner, dd', be turned up towards the eye, it is plain that we should see the upper face in addition to the two faces seen in the last elevation, Pl. IX., Fig. 99. It is also plain that the cube may thus be turned up just far enough to make these three visible faces appear equal, as in Pl. IX., Fig. 100. Then the edge, c—c', Fig. 99, will appear equal to Ce'' and Cf'', Fig. 100.

200. Let us now observe carefully the facts which result when the corner dd' is turned up just far enough to make the three visible faces of the cube appear equal.

1st. When the faces of the cube appear equal, the sides of those faces, i. e., the edges of the cube must appear equal; hence a—a', b—b', c—c', Fig. 99, for example, will be equal in the new projection.

2nd. Being equal, both in projection and in space, these edges must be equally inclined to the plane of projection (198).

3rd. These edges being equal, equally inclined to the plane of projection, and also equally inclined to each other—being at right angles to each other in space—their included angles will be equal in projection. (See Pl. III., Fig. 35, the plan.) Hence, as a first step in constructing the new projection, lay off from any assumed point, C, Pl. IX., Fig. 100, three lines making angles of 120° with each other; observing that if one, Cn, be vertical, the other two, Cf'' and Ce'', will make angles of 30° with a horizontal line through C, and can therefore be easily drawn with the 30° triangle.

4th. The diagonal ef—$e'f'$ of the upper base, Pl. IX., Fig. 99, remaining parallel to the vertical plane, will still be seen in its true size; hence from C, Pl. IX., Fig. 100, lay off half of the diagonal ef—$e'f'$ from Fig. 99, each way on the line EF, and at E and F erect perpendiculars, whose intersection with Ce'' and Cf'' will give the required diagonal in its true position, and consequently Ce'' and Cf'', two sides of the required projection, and projections of ce—$d'e'$ and cf—$d'f'$, Fig. 99. Make now $Cn = Ce''$.

5th. All the other edges of the cube are equal and parallel to Ce'', Cf'', and Cn. They are therefore readily drawn, and the required projection completed. This projection is called the Isometrical Projection of the cube.

6th. We see now, by inspection, that the perimeter, D$f''onpe''$, of this projection is a regular hexagon, of whose circumscribing circle, C is the centre, and Ce'', Cf'', and Cn, are equidistant radii. It also appears that Fig. 100 could have been constructed by taking $e''f''=ef$ from Fig. 99, as the side of an inscribed equilateral triangle, by then inscribing a hexagon in the circumscribing circle of that triangle, and by then drawing Ce'', Cf'', and Cn. Once more, it appears that the three diagonals, $e''f''$, $f''n$, and ne'' are equal, and that the diagonal of the body of the cube, commencing at C, is projected in the point C. But when a straight line is projected in a point, it is perpendicular to the plane of projection. Hence in Isometrical projection, the plane of projection is perpendicular to a diagonal of the body of the cube.

201. Summing up in the form of definitions:

1st. Isometrical projection is that in which the three plane right angles, which form a solid right angle, appear equal in projection.

2nd. The three lines meeting at C are called isometric axes.

3rd. The point C is called the isometric centre.

4th. Planes containing any two of the isometric axes, or parallel to these axes, are called isometric planes.

5th. Lines parallel to the isometric axes are called isometric lines.

202. It is to be noticed, that while it is the three diagonals only of the faces of the cube, Fig. 100, that are seen in their real size, or on the same scale as that used for the ordinary plan and elevation, yet the edges are the lines of which the measurements are usually given, and which it is therefore desirable to represent on the same scale in all its projections. If, therefore, a figure be constructed, as in Pl. IX., Fig. 101, similar to Fig. 100, but having its edges, Da, &c., equal to the edges of the plan in Fig. 99, such a figure will be called an *Isometrical Drawing* of the cube shown in Fig. 99, while it would be the strict isometrical *projection* of a larger and imaginary cube.

This imaginary cube is to Fig. 101, as Fig. 99 is to its projection in Fig. 100.

203. Again, it is evident that it is the *direction* of the projection of the edges of a solid right angle which determines whether the projection is isometrical or not; hence, not merely cubes, but any

rectangular objects, can readily be put into isometrical projection.

204. Recollecting that all lines of the object in space which are at right angles to each other, are shown in the isometrical *drawing* on the scale of the plan and elevation, and recollecting that in most pieces of carpentry, in boxes, furniture, &c., the principal lines *are* at right angles to each other, we see at once the great convenience of isometrical drawing, which shows the true dimensions of objects in three directions on the same figure, and gives a clearer idea of the construction of objects than plans and elevations do. The disagreeably distorted appearance, however, of large or complicated structures, as buildings or machines, when shown in isometrical projection, causes this method to be confined, in its application, to small objects, details, &c.

205. We know that we can find the projection of any line, straight or curved, by finding the projections of a certain number of its points. If, then, we can find the isometric projection of a point in any situation, we shall then be able to find the isometric projection of any line, surface, or solid.

Let us therefore take the general Problem in its three cases:

1st. To find the isometric projection of a point in one of the edges of a square right prism.

2nd. Ditto of a point in one of the faces of the prism.

3rd. Ditto of a point anywhere in the interior of the prism.

CHAPTER II.

PROBLEMS INVOLVING ONLY ISOMETRIC LINES.

206. PROB. 1. *To make the isometrical drawing of a cube, from a given plan and elevation.*

1st Method. The plan and elevation, Pl. IX., Fig. 98, represents a cube, whose edge is one inch in length, seen in "*right projection.*" To make the isometrical *drawing* of the same cube, Fig. 101, take any point, C, as a centre, and describe a circle with one inch radius, CD ; lay off the radius six times, commencing at either end of the vertical diameter; join the points thus formed, and draw aC, bC, and cC. Then Ca and Cb, and all the lines parallel with them, are drawn with the 30° triangle.

2nd Method. Draw the three axes, and lay off an inch on each of them. The remaining lines, being equal and parallel to these, are drawn with the 30° triangle, placed in different positions, which will readily suggest themselves, on attempting the construction.

207. To complete the representation, certain lines are made heavy. If a line, LL′, Pl. IX., Fig. 102, be drawn through the corner, a, of the cube, in the direction of a ray of light, it will pass through the cube and leave it at the corner, p. Those surfaces of the cube, therefore, which are represented in Fig. 102 by CqDa, aCno, and aDCo, are illuminated, and equally so, because the diagonal, as ap, makes equal angles with all the faces of the cube. But Cqpn, CpqD, and Cpno, are all in the dark; hence, according to the usual rule, that those lines which divide a dark from a light surface are made heavy, Dq, qC, Cn, and no, are all made heavy.

Again, we know from elementary geometry that two intersecting lines fix the position of the plane containing them. Accordingly, one plane can contain oab and the ray LL′. This plane will also contain pq, parallel to oab, and hence will cut aDqC in aq.

Therefore, rays through all points of ab will meet aDqC in aq. Hence, by drawing the ray bb′, we find ab′, the shadow of ab, which is the edge oa produced.

By similar reasoning, we find ad′, the shadow of ad, which is the edge Da produced.

208. PROB. 2. *To represent a prismatic block, as cut from a*

corner of the cube. Pl. IX., Fig. 101. This problem involves the construction of points in the isometric axes. Let the cube now be supposed to be one whose edge is five inches long, and let it be drawn to a scale of one fifth, or of one inch to five inches. Suppose a piece, each of whose sides is a rectangle, to be cut from the corner nearest the eye. Let $Ca'=2$ inches, $Cb'=3$ inches, and $Cc'=$ 1 inch. Lay off these distances from C, upon the axes, and through the points a', b', and c', draw isometric lines, as $a'd$ and $b'd$, which, by their intersections, will complete the portion cut from the cube.

Heavy lines are placed as in the last problem; but all the lines of division between the body of the cube and the piece cut from it are light, according to the common rule, that when two surfaces are continuous and form one surface, the dividing line between them is fine; and this, because such a dividing line does not separate a light from a dark surface, but merely lies in a light surface.

209. PROB. 3. *To construct the isometrical drawing of a carpenter's oil stone box.* Pl. X., Fig. 103. This problem involves the finding of points which are in the planes of the isometric axes.

Let the box containing the stone be 10 inches long, 4 inches wide, and 3 inches high, and let it be drawn on a scale of $\frac{1}{4}$.

Assume C then, and make $Ca=4$ inches, $Cb=10$ inches, and $Cc=3$ inches, and by other lines aD, Db, &c., equal and parallel to these, complete the outline of the box.

Represent the joint between the cover and the box as being 1 inch below the top, aCb. Do this by making $Cp=1$ inch, and through p drawing the isometric lines which represent the joint.

Suppose, now, a piece of ivory 5 inches long and 1 inch wide to be inlaid in the longer side of the box. Bisect the lower edge at d, make $de=1$ inch, and $ee'=1$ inch. Through e and e', draw the top and bottom lines of the ivory, making them $2\frac{1}{2}$ inches long on each side of de', as at $e't$, $e't'$. Draw the vertical lines at t and t', which will complete the ivory.

210. To show another way of representing a similar inlaid piece, let us suppose one to be in the top of the cover, 5 inches long and $1\frac{1}{2}$ inches wide. Draw the diagonals ab and CD, and through their intersection, o, draw isometric lines; lay off $oo'=2\frac{1}{2}$ inches, and $oo''=\frac{3}{4}$ of an inch, and lay off equal distances in the opposite directions on these centre lines.

Through o', o'', &c., draw isometric lines to complete the representation of the inlaid piece.

211. PROB. 4. *To represent the same box (Prob. 3), with the cover removed.* Pl. X., Fig. 104.

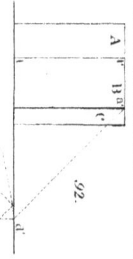

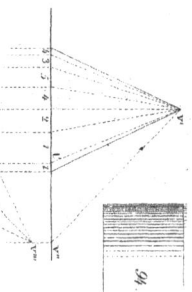

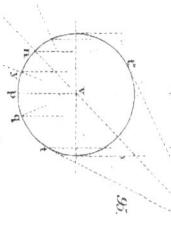

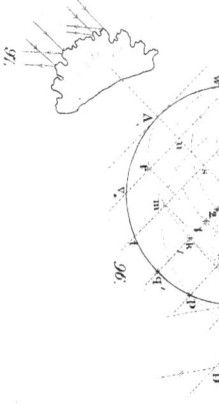

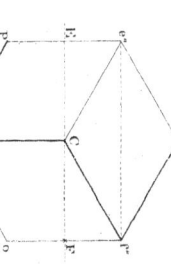

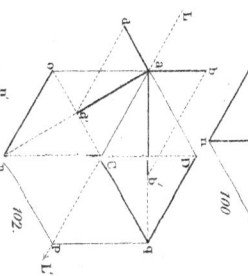

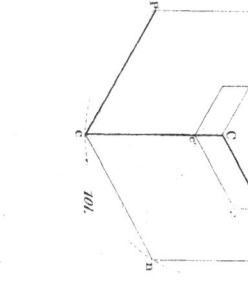

This problem involves the finding of the positions of points not in the isometric planes.

Supposing the edges of the box to be indicated by the same letters as are seen in Fig. 103, and supposing the body of the box to be drawn, 10 inches long, 4 inches wide, and 2¼ inches high; then, to find the nearest upper corner of the oil stone, lay off on C*b* 1½ inches, and through the point *f*, thus found, draw a line, *ff'*, parallel to C*a*. On *ff'*, lay off 1 inch at each end, and from the points *hh'*, thus found, erect perpendiculars, as *hn*, each ¾ of an inch long. Make the further end, *x*, of the oil stone 1½ inches from the further end of the box, and then complete the oil stone as shown. To find the panel in the side of the box, lay off 2 inches from each end of the box, on its lower edge; at the points thus found, erect perpendiculars, of ¾ of an inch in length, to the lower corners, as *p*, of the panel; make the panels 1½ inches wide, as at *pp'*, and ½ an inch deep, as at *pr*, which last line is parallel to C*a;* and, with the isometric lines through *r*, *p*, *p'*, and *t*, completes the panel.

212. *The manner of shading this figure* will now be explained.

The top of the box and stone is lightest. Their ends are a trifle darker, since they receive less of the light which is diffused through the atmosphere. The shadow of the oil stone on the top of the box is much darker than the surfaces just mentioned. The shadow of the foremost vertical edge of the stone is found in precisely the same way as was the shadow of the wire upon the top of the cube, Pl. IX., Fig. 102. The sides of the oil stone and of the box, which are in the dark, are a little darker than the shadow, and all the surfaces of the panel are of equal darkness and a trifle darker than the other dark surfaces. In order to distinguish the separate faces of the panel, when they are of the same darkness, leave their edges very light. The little light which those edges receive is mostly perpendicular to them, regarding them as rounded and polished by use. These light lines are left by tinting each surface of the panel, separately, with a small brush, leaving the blank edges, which may, if necessary, be afterwards made perfectly straight by inking them with a light tint. The upper and left hand edges of the panel, and all the lines corresponding to those which are heavy in the previous figures, may be ruled with a dark tint. In the absence of an engraved copy, the figures will indicate tolerably the relative darkness of the different surfaces; 1 being the lightest, and the numbers not being consecutive, so that they may assist in denoting *relative* differences of tint. When this figure is thus shaded, its edges should not be inked with ruled lines in black ink, but should be inked with pale ink

CHAPTER III.

213. WHEN non-isometrical lines occur, each point in any of them which is in an isometric plane, must be located by two isometric lines representing lines which in common projection would be found at right angles to each other, as is illustrated in the following problems—which, by the way, are numbered continuously through the chapters of this DIVISION.

214. PROB. 5. *To construct the isometrical drawing of the scarfed splice, shown at* Pl. VII., Fig. 66. Let the scale be $\frac{1}{16}$, or three fourths of an inch to a foot. In this case, Pl. X., Fig. 105, it will be necessary to reconstruct a portion of the elevation to the new scale (see Pl. X., Fig. 106), where $Ap=1\frac{1}{2}$ feet, $pA''=6$ inches, and the proportions and arrangement of parts are like Pl. VII., Fig. 66. In Fig. 105, draw AD, 3 feet; make DB and AA' each one foot, and draw through A' and B isometric lines parallel to AD. Join A—B, and from A lay off distances to 1, 2, 3, 4, equal to the corresponding distances on Fig. 106. Also lay off, on BK, the same distances from B, and at the points thus found on the edges of the timber, draw vertical isometric lines equal in length to those which locate the corners of the key in Fig. 106. Notice that opposite sides of the keys are parallel, and that AV, and its parallel at B, are both parallel to those sides of the keys which are in space perpendicular to AB. To represent the obtuse end of the upper timber of the splice, bisect AA', and make $va=Aa$, Fig. 106, and draw Aa and $A'a$. Locate m by va produced, as at ar, Fig. 106, and a short perpendicular$=rm$, Fig. 106, and draw mV and mV'; VV' being parallel to AA', and am being parallel to AV. To represent the washer, nut, and bolt, draw a centre line, vv', and at t, the middle point of Vn, draw the isometric lines tu and ue, which will give e, the centre of the bolt hole or of the bottom of the washer. A point —coinciding in the drawing with the upper front corner of the nut— is the centre of the top of the washer, which may be made $\frac{3}{4}$ of an inch thick.

Through the above point draw isometric lines, rr' and pp', and lay off on them, from the same point, the radius of the washer, say $2\frac{1}{2}$ inches, giving four points, as o, through which, if an isometric square be drawn, the top circle of the washer can be sketched in it, being tangent to the sides of this square at the points, as o, and elliptical (oval) in shape. The bottom circle of the washer is seen throughout a semi-circumference, i. e. till limited by vertical tangents to the upper curve.

On the same centre lines, lay off from their intersection, the half side of the nut, $1\frac{1}{4}$ inches, and from the three corners which will be visible when the nut is drawn, lay off on vertical lines its thickness, $1\frac{1}{4}$ inches, giving the upper corners, of which c is one. So much being done, the nut is easily finished, and the little fragment of bolt projecting through it can be sketched in.

The other nut, being similarly constructed, is omitted. The nut is here constructed in the simplest position, i. e. with its sides in the direction of isometric lines. If it had been determined to construct it in any oblique position, it would have been necessary to have constructed a portion of the plan of the timber with a plan of the nut—then to have circumscribed the plan of the nut·by a square, parallel to the sides of the timber—then to have located the corners of the nut in the sides of the isometric drawing of the circumscribed square. Let the student draw the other nut on a large scale and in some such irregular position. See Fig. 107, where the upper figure is the isometrical drawing of a square, as the top of a nut; this nut having its sides oblique to the edges of the timber, which are supposed to be parallel to ca.

215. Prob. 6. *To make an isometrical drawing of an oblique timber framed into a horizontal one.* Pl. X., Fig. 108. Let the original be a model in which the horizontal piece is one inch square, and let the scale be $\frac{1}{4}$.

Make ag and ah, each one inch, complete the isometric end of the horizontal piece, and draw ad, hn, and gk. On ad, lay off $ab=2$ inches, and draw bm, $bc=2\frac{1}{4}$ inches, and $cd=2$ inches. Make $de=\frac{3}{4}$ of an inch, draw ce, and lines through b and m parallel to it. Let f be $\frac{1}{4}$ of an inch below ad and $2\frac{1}{8}$ inches from ag, measured from a on ad, and draw bf and fc, which, with the representation of the broken ends, will complete the figure.

216. Prob. 7. *To make an isometrical drawing of a pyramid standing upon a recessed pedestal.* Pl. X., Fig. 109. (From a

Model.) Let the scale be ½. Assuming C, construct the isometric
square, CABD, of which each side is 4¼ inches. From each of its
corners lay off on each adjacent side 1¼ inches, giving points, as *e*
and *f;* from all of these points lay off, on isometric lines, distances
of ¾ of an inch, giving points, as *g, k,* and *h.* From all these points
now found in the upper surface, through which vertical lines can be
seen, draw such lines, and make each of them one inch in length,
and join their lower extremities by lines parallel to the edges of the
top surface.

Isometric lines through *g, m,* &c., give, by their intersections, the
corners of the base of the pyramid, that being in accordance with
the construction of this model, and the intersection of AD and CD
is *o,* the centre of that base. The height of the pyramid—1⅝ inches
—is laid off at *ov.* Join *v* with the corners of the base, and the
construction will be complete.

217. *To find the shadows on this model.*—According to principles enunciated in DIVISION III., the shadow of *fh* begins at *h,*
and will be limited by the line *hs, s* being the shadow of *f,* and the
intersection of the ray *fs* with *hs.*   *st* is the shadow of *f*F. According to (207), *w* is the shadow of *v,* and joining *w* with *p* and *q,* the
opposite corners of the base, gives the boundary of the shadow on
the pedestal. *ka* is the boundary of the shadow of *gk* on the face
ku. The heavy lines are as seen in the figure. If this drawing is
to be shaded, the numerals will indicate the darkness of the tint for
the several surfaces—1 being the lightest.

218. PROB. 8. *To construct the isometrical drawing of a wall in*
batter, with counterforts, and the shadows on the wall. Pl. X.,
Fig. 110. A wall in batter is a wall whose face is a little inclined
to a vertical plane through its lower edge, or through any horizontal line in its face. Counterforts, or buttresses, are projecting parts
attached to the wall in order to strengthen it. Let the scale be
one fifth, i. e., let each one of the larger spaces on the scale marked
50, on the ivory scale, be taken as an inch.

Assume C, and make CD=2 inches and CI=10 inches in the right
and left hand isometrical directions. Make DF=6 inches, EF=1½
inches, FG=10 inches, and GH=1½ inches, and join H and E, all
of these being isometrical lines. Next draw EC, then as the top
of the counterfort is one inch, vertically, below the top of the wall,
while EC is not a vertical line, make F*d*=one inch, draw *de* parallel to FE and *ef,* parallel to EH. Make *ef* and C*a* each one inch,
at *f* make the isometric line *fg*=¾ of an inch, and at *a* make *ab*=2

inches, and draw *bg*.   Make *bc*=1½ inches, draw *ch* parallel to *bg* and complete the top of the counterfort.   The other counterfort is similar in shape and similarly situated, i. e., its furthermost lower corner, *k*, is one inch from I on the line IC, while HI is equal and parallel to CE.

219. *To find the shadows of the counterforts on the face of the wall.*—We have seen—Pl. VIII., Fig. 78—that when a line is perpendicular to a vertical plane, its shadow on that plane is in the direction of the projection of the light upon the same plane, and from Pl. IX., Fig. 102, that the projection, *an*, of a ray on the left hand vertical face of a cube makes on the isometrical drawing an angle of 60° with the horizontal line *nn'*.

Now to find the shadow of *mn*.   The front face of the wall—Pl. X., Fig. 110—not being vertical, drop a perpendicular, *fo*, from an upper back corner of one of the counterforts, upon the edge *ba*, produced, of its base, and through the point *o*, thus found, draw a line parallel to CI.   *nfop* is then a vertical plane, pierced by *mn*, an edge of the further counterfort which casts a shadow; and *np* is the direction of the shadow of *mn* on this vertical plane. The shadow of *mn* on the plane of the lower base of the wall is of course parallel to *mn*, and *p* is one point of this shadow, hence *pq* is the direction of this shadow.   Now *n*, where *mn* pierces the actual face of the wall, is one point of its shadow on that face, and *q*, where its shadow on the horizontal plane pierces the same face, is another point, hence *nq* is the general direction of the shadow of *mn* on the front of the wall, and the actual extent of this shadow is *nr*, *r* being where the ray *mr* pierces the front of the wall.

From *r*, the real shadow is cast by the edge, *mu*, of the counterfort.   *r*, the shadow of *m*, is one point of this shadow, and *s*, where *um* produced, meets *yn* produced, is another (in the shadow produced), since *s* is, by this construction, the point where *um*, the line casting the shadow, pierces the surface receiving the shadow. Hence draw *srt*, and *nrt* is the complete boundary of the shadow sought.   The shadow of the hither counterfort is similar, so far as it falls on the face of the wall.

Other methods of constructing this shadow may be devised by the student.   Let *t* be found by means of an auxiliary shadow of *um* on the plane of the base of the wall.

CHAPTER IV.

220. Prob. 9. *To make an exact construction of the isometrical, drawing of a circle.* Pl. X., Figs. 111–112. This construction is only a special application of the general problem requiring the construction of points in the isometric planes.

Let Pl. X., Fig. 111, be a square in which a circle is circumscribed. The rhombus—Fig. 112—is the isometrical drawing of the same square, CA being equal to $C'A'$. The diameters $g'h'$ and $e'f'$ are those which are shown in their real size at gh and ef, giving g, h, e, and f as four points of the isometrical drawing of the circle. In Fig. 111, draw $b'a'$, from the intersection, b', of the circle with $A'D'$ and parallel to $C'A'$. As a line equal to $b'a'$, and a distance equal to $A'a'$ can be found at each corner of Fig. 111, lay off each way from each corner of Fig. 112, a distance, as Aa, equal to $A'a'$, and draw a line ab parallel to CA and note the point b, where it meets AD. Similarly the points n, o, and r may be found. Having now eight points of the ellipse which will be the isometrical drawing of the circle, and knowing as further guides, that the curve is tangent to the circumscribing rhombus at g, h, e and f, and perpendicular to its axes at b, n, o and r, this ellipse can be sketched in by hand, or by an irregular curve.

221. If, on account of the size of the figure, more points are desirable they can readily be found. Thus; on any side of Fig. 111, take a distance as $C'c'$ and $c'd'$ perpendicular to it, and meeting the circle at d'. In Fig. 112, make Cc=$C'c'$, and make cd equal to $c'd'$ and parallel to CD, then will d be the isometrical position of the point d'.

222. Prob. 10. *To make an approximate construction of the isometrical drawing of a circle.* Pl. X., Fig. 113. By trial we shall find that an arc, gf, having C for a centre and Cf for its radius, will very nearly pass through n; likewise that an arc eh, with B for a centre, will very nearly pass through r. These arcs will be tangents to the

sides of the circumscribing square at their middle points, as they should be, since Cf and Be are perpendicular to these sides at their middle points. Now in order that the small arcs, fbh and goe, should be both tangent to the former arcs and to the lines of the square at g, h, f and e, their centres must be in the radii of the larger arcs, hence at their intersections p and q. Arcs having p and q for centres, and ph and qe as radii, will complete a four-centred curve which will be a sufficiently near approximation to the isometrical ellipse, when the figure is not very large, or when the object for which it is drawn does not require it to be very exact.

223. PROB. 11. *To make an isometrical drawing of a solid composed of a short cylinder capped by a hemisphere.* Pl. X., Fig. 114. Scale$=\frac{1}{4}$. Let this body be placed with its circular base lowermost, as shown in the figure. Make ac and bd the height of the cylindrical part$=1\frac{5}{12}$ inches, and draw cd. Now a sphere, however looked at, must appear as a sphere, hence take e, the middle point of cd, as a centre, and ec as a radius, and describe the semicircle ehd, which will complete the figure.

224. In respect to execution, in general, of the problems of this DIVISION, a description of it is not formally distinguished from that of their construction, since the figures generally explain themselves in this respect. In the present instance, the visible portion of the only heavy line required will be the arc anb. As there is no angle at the union of the hemisphere with the cylinder—see the preceding problem—no full line should be shown there, but a dotted curve parallel to the base and passing through c and d, might be added to show the precise limits of the cylindrical part.

225. Again, if it be desired to shade this body, the element, ny, of the cylindrical part, with the curve of shade, $pfry$, on the spherical part, will constitute the darkest line of the shading. The curve of shade, $pfry$, is found approximately as follows. The line ny being the foremost element of the cylinder, $yy'=ne$, is the projection of an actual diameter of the hemisphere. mm'' and gg'', parallel to ST, are the radii of small semicircles of the hemisphere, to which projections of rays of light may be drawn tangent, and m' and g', are the true positions of their centres—ym' being equal to nm, and yg' equal to ng. Drawing arcs of such semicircles, and drawing rays, fd and rS, tangent to them, we determine f and r, points of the curve of shade on the spherical part of the body, through which, with p and y, the curve may be sketched.

226. PROB. 12. *To construct the isometrical circles on the three*

visible faces of a cube, as seen in an isometrical drawing. Pl. XI., Fig. 115. This figure needs no minute description here, being given to enable the student to become familiar with the position of isometrical circles in the three isometrical planes, and with the positions of the centres used in the approximate construction of those circles. By inspection of the figure, the following general principle may be deduced. The centres of the larger arcs are always in the obtuse angles of the rhombuses which represent the sides of the cube, and the centres of the smaller arcs are at the intersection of the radii of the larger arcs with the diagonals joining the acute angles of the same rhombuses—i. e. the longer diagonals.

227. PROB. 13. *To make the isometrical drawing of a bird house.* Pl. XI., Fig. 116. Assuming C, make CA′=16 inches, Ca=3 inches, and CB=9 inches. At a, make ab=1 inch, and ac=8 inches. Draw next the isometric lines BD and cD. Through b make bE=16 inches, make EA=3 inches, and EF=7 inches. Then draw the isometric lines DH and FH. Bisect bE at N, make Nf=11½ inches, draw cf and Ff, make ce=Fh=fg=one inch, and draw eg and hg. Through A and b draw isometric lines which will meet, as at $a′$. On bE make bk, lm and nE each equal to 3 inches, and let kl and mn each be 3½ inches. At l and n draw lines, as lv, parallel to CB, and one inch long, and at their inner extremities erect perpendiculars, each 3½ inches long. Also at k, l, m and n, draw vertical isometrical lines, as kt, 3½ inches long. The rectangular openings thus formed are to be completed with semicircles whose real radius is 1¾ inches, hence produce the lines, as kt—on both windows—making lines, as kG, 5¼ inches long, and join their upper extremities as at GI. The horizontal lines, as ts, give a centre, as s, for a larger arc, as tu. The intersection of Go with Iz—see the same letters on Fig. 115—gives the centre, p, of the small arc, uo. The same operations on both openings make their front edges complete. Make oq and pr parallel, and each, one inch long, and r will be the centre of a small arc from q which forms the visible part of the inner edge of the window. Suppose the corners of the platform to be rounded by quadrants whose real radius is 1½ inches. The lines $a′b$ and bk each being 3 inches, k is the centre for the arc which represents the isometric drawing of this quadrant, whose real centre on the object, is indicated on the drawing at y. So, near A, w is the centre used in drawing an arc, which represents a quadrant whose centre is x.—See the same letters on Fig. 115.

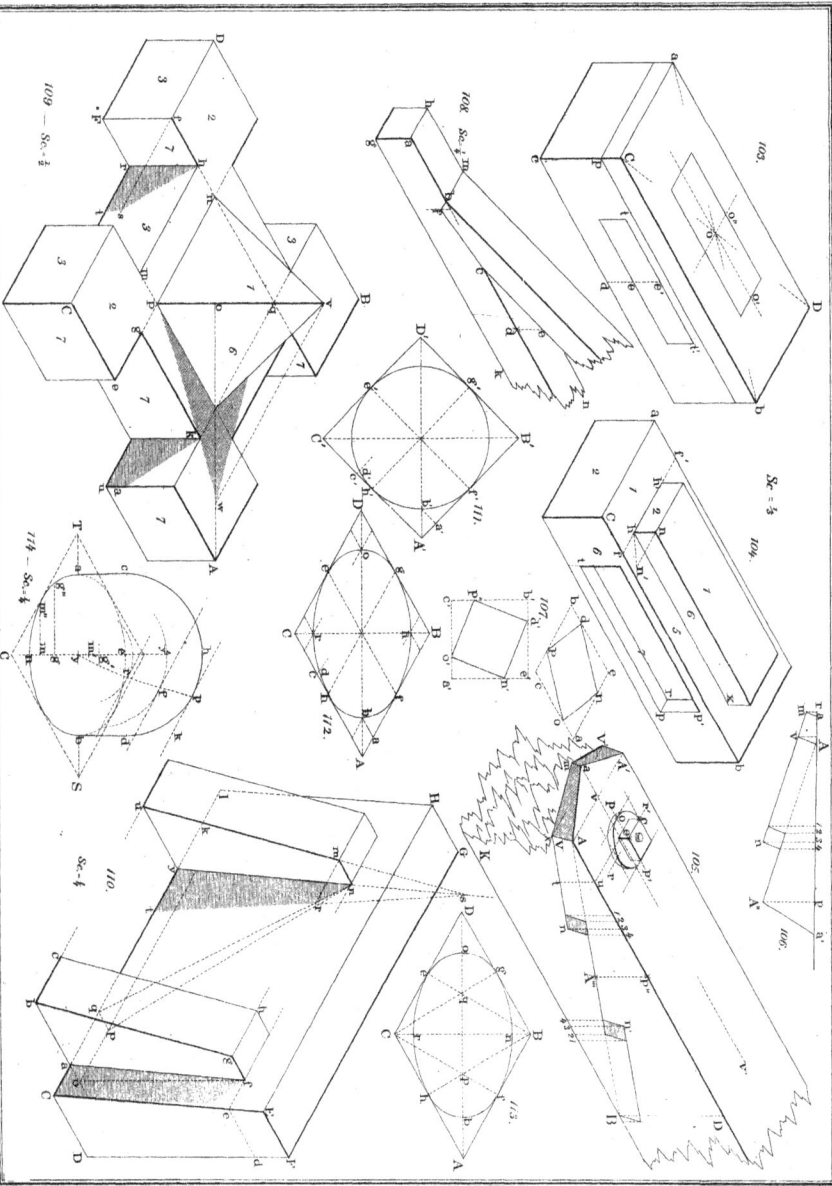

Of the Isometrical Drawing of Circles which are divided in Equal Parts.

228. PROB. 14. Pl. XI., Fig. 117. *First method.*—If the semi-ellipse, ADB, be revolved up into a vertical position about AB as an axis, it will appear as a semicircle AD'B of which ADB is the isometrical projection. Since AB, the axis, is parallel to the vertical plane, the arc in which any point, as D, revolves, is in a plane perpendicular to the vertical plane, and is therefore projected in a straight line DD'. Hence to divide the semi-ellipse ADB into parts corresponding to the parts of the circle which it represents, divide AD'B into the required number of equal parts, and through the points thus found, draw lines parallel to D'D, and they will divide ADB in the manner required. The opposite half of the curve can of course be divided in a similar manner.

229. *Second method.*—CE is the true diameter of the circle of which ADB is the isometrical drawing. Let it also represent the side of the square in which the original circle to be drawn is inscribed. The centre of this circle is in the centre of the square, hence at O, found by making *e*O equal to half of CE, and perpendicular to that line at its middle point *e*.

With O as a centre, draw a quarter circle, limited by CO and EO, and divide it into the required number of parts. Through the points of division, draw radii and produce them till they meet CE. CE, considered as the side of the isometrical drawing of the square, is the drawing of the original side CE of the square itself with all its points 1, 2, ... 6, 7, &c., and O' is the isometrical position of O. Hence connect the points on CE with the point O' and the lines thus made will divide the quadrant BC' in the manner required.

Applications of the preceding Problem.

230. PROB. 15. *To make an isometric drawing of a segment of an Ionic Column.* Pl. XI., Fig. 118. Let *a*D be a side of the circumscribing prism of the column. By the second method of Prob. 14, find O', the centre of a section of the column, and with O' as a centre, draw any arc, as *a'q'*. The curved recesses in the surface of a column are called flutes, or the column is said to be fluted. In an Ionic, and in some other styles of columns, the flutings are semicircular with narrow flat, or strictly, cylindrical surfaces, as *ee"p*, between them. Hence, in Fig. 118, assume *a'b'*, equal to *q'v'*, as half of a space between two flutes, divide *b'v'* into four equal parts, and make the points of division central points of the spaces as *f"e'*

6

between the flutes. Let the flutes be drawn with points, as c' as centres and touching the points as $b'd'$; then draw an arc tangent, as at r, to the flutes. To proceed now with the isometrical drawing, draw, in the usual way, the isometrical drawing of the outer circumferences of the column, tangent to aD and b'''F—assuming DF for the thickness of the segment. Now $a'q'$ being any arc, and not one tangent to aD so as to represent the true size of a quadrant of the outer circumference, the true radius of the circle tangent to the inner points of all the flutes will be a fourth proportional, $O'y'$, to $O'f'$, Oi ($=O'y$), and $O's$. On Oi, lay off $OY=Oy'$, draw IJ to find a centre I, and similarly find the other centres of the larger arcs of the inner ellipse. The points n, h and n', h' are the centres of the small arcs (222) for the two bases. Having gone·thus far, produce $O'b'$, $O'c'$, &c. to aD ; at b, c, &c., erect vertical lines, bb''', cc''', &c., then from b, c, &c. draw lines to O, and note their intersections, b'', c'', &c. with the curves of the lower base; and from b''', c''', &c. draw lines to O″ and note their intersections, b'''', c'''', &c. with the ellipses of the upper base. This process gives three points for each flute by which they can be accurately sketched in, remembering that they are tangent to the inner dotted ellipses, as at c'''', o''', &c. and to the radii, as e''O″—at e''. Parts beyond FO″ are projected over from the parts this side, thus drawn.

231. PROB. 16. *To construct the isometrical drawing of a segment of a Doric Column.* Pl. XI., Fig. 119. The flutes of a Doric column are shallow and have no flat space between them. Adopting the first method of Prob. 14, let the centre, A, of the plan be in the vertical axis, GA′, of the elevation, produced. Let Ac and Ab be the outer and inner radii containing points of the flutes. Make A$d=\frac{4}{3}$ of Ac, for the radius of the circle which shall contain the centres of the flute arcs. Let there be four flutes in the quadrant, shown in the plan. Their centres will be at h, &c., where radii Ag, &c., bisecting the flutes, meet the outermost arc. In proceeding to construct the isometrical drawing, project b and c, at b' and c' on the axis A′d'. Now, owing to the variation at b and c' between the true and the approximate ellipse, we cannot make use of the latter, if we retain b' and c' in their proper places, as projected from b and c, hence through b' and c' draw isometric lines which locate the points N′ and Q′ (the points are between these letters) which are the true positions of N and Q respectively. Corresponding points, between N‴ and v, are similarly found. By an irregular curve the semi-ellipses vb'Q′ and N‴c'N′ can be quite accurately

drawn. Next, project upon these curves the points u, e, &c., r, g, &c. of the flutes—as at u', e', &c., r', &c., and with an irregular curve draw the curves through these points, tangent to the inner semi-ellipse. The corresponding curves of the lower base are found by drawing lines $r'r''$, $u'u''$, &c. through the points of tangency, r', k', &c., and through u', &c., and all equal to FD, the thickness of the segment.

The curves above the axis $A'd'$ are projected across from those already made below it.

Special Examples.

232. PROB. 17. *To draw a cube or other parallelopipedical body so as to show its under side.* Pl. XI., Fig. 120. By reflection, it becomes evident that it is the relative direction of the lines of the drawing among themselves, that make it an isometrical drawing. Hence in the figure, where all the lines are isometric lines, the whole is an isometric drawing, now that the solid angle C is nearest us, as much as if the angle A (lettered C on previous figures) were nearest us.

233. *Remark.* By a curious exercise of the will, we can make Fig. 120 appear as an interior view, showing a floor CFED, and two walls; or, in Fig. 115 and others, we can picture to ourselves an interior showing a ceiling GIk and two walls. This is probably because—1*st.* All drawings being of themselves only plane figures, we educate the eye to see in them, what the mind chooses to conceive of, as having three dimensions. 2*nd.* When, as in isometrical drawing, the drawing in itself as a plane figure, is the same for an interior as for an exterior view of any given magnitude, the eye sees in it whichever of these two the mind chooses to imagine.

234. PROB. 18. *To construct isometrical drawings of oblique sections of a right cylinder with a circular base.* Pl. XI., Fig. 121. This construction is easily made from a given circle as a base of the cylinder, that base being in an isometric plane. The circle in the plane AGEF is such a circle. Let A'G'E'F' be a plane inclined to AGEF but perpendicular, as the latter is, to the planes GB and DF, and let A''G''E''F'' be a plane inclined to all the sides of the prism AGE—D.

Lines, as $aa'a''$, &c., being in the faces of the prism and parallel to their edges, meet the intersections, F'E'—F''E'', &c. of the oblique planes at points a', a'', &c., which are points of oblique sections of a cylinder inscribed in the prism AGE—D, and whose base is $acbdu$,

So, points, as c, have the corresponding points $c'c''$, &c. in the diagonals A'E', A''E'' of the planes in which those points are found. To find points, as t', t'', &c. corresponding to t in the base, draw any line, as yd, through t, and find the corresponding lines, as $y'd'$ and $y''d''$. Their intersections with the diagonals G'F' and G''F'' will give the points t', t'', &c. Having thus found eight points of each oblique section of the given inscribed cylinder whose base is $abcd$–u, and remembering that each of these sections is tangent to the sides of its circumscribing polygon (considering the lines $y'd'$,&c.), the curves a', b', c', t', and a'', b'', c'', t'' are readily sketched in.

235. *Remarks.* *a.* As before stated, it is the relative direction, among themselves, of the lines of an isometrical drawing, that determine it as an isometrical drawing, hence Pl. XI., Fig. 121, is an isometrical drawing, though its lines are not situated with reference to the edges of the plate as the similar lines of previous figures have been. If the portion of the plate containing this figure were cut out so as to make the edges of the fragment, so cut out, parallel and perpendicular to GE, the figure would appear like the previous isometrical drawings.

b. The problem just solved must not be confounded with one which should seek *to find the isometric projection of a curve which in space is a circle on the plane* G'E'—A', for the curve $a'b'c'd't'$ is not a circle, in space.

236. PROB. 19. *To solve the problem just enunciated.* Pl. XI., Figs. 121–122. $e''r$—Fig. 122—is a plan of the section rA'F' in which—it being a square—a circle can be inscribed. $e''r$ is therefore the plan of the circle also. Making rG—Fig. 122—equal to rG'—Fig. 121, and drawing e''G, we have the plan of the section G'E'—A', and making $o''p'$, Fig. 122, equal to $e''r$, we have the plan of a circle in the section G'E'—A'. Now draw $o''x$ and $p'e'$—Fig. 122—make A'e and G'e''' and e'''p and eo—Fig. 121—equal to $e'''x$, Ge', $e'p'$ and xo''—Fig. 122—draw pY and oU; and u'U and b'Y to intersect them, and we shall have U and Y as the isometric positions in the plane G'E'—A' of the points o' and p' which, considered as points on the circle, are evidently enough extremities of its horizontal diameter, at which points, the circle is tangent to the vertical lines whose isometric positions in the plane G'E'—A' are pY and oU. T and a' are other points.

The finding of intermediate points, which is not difficult, is left as an exercise for the student.

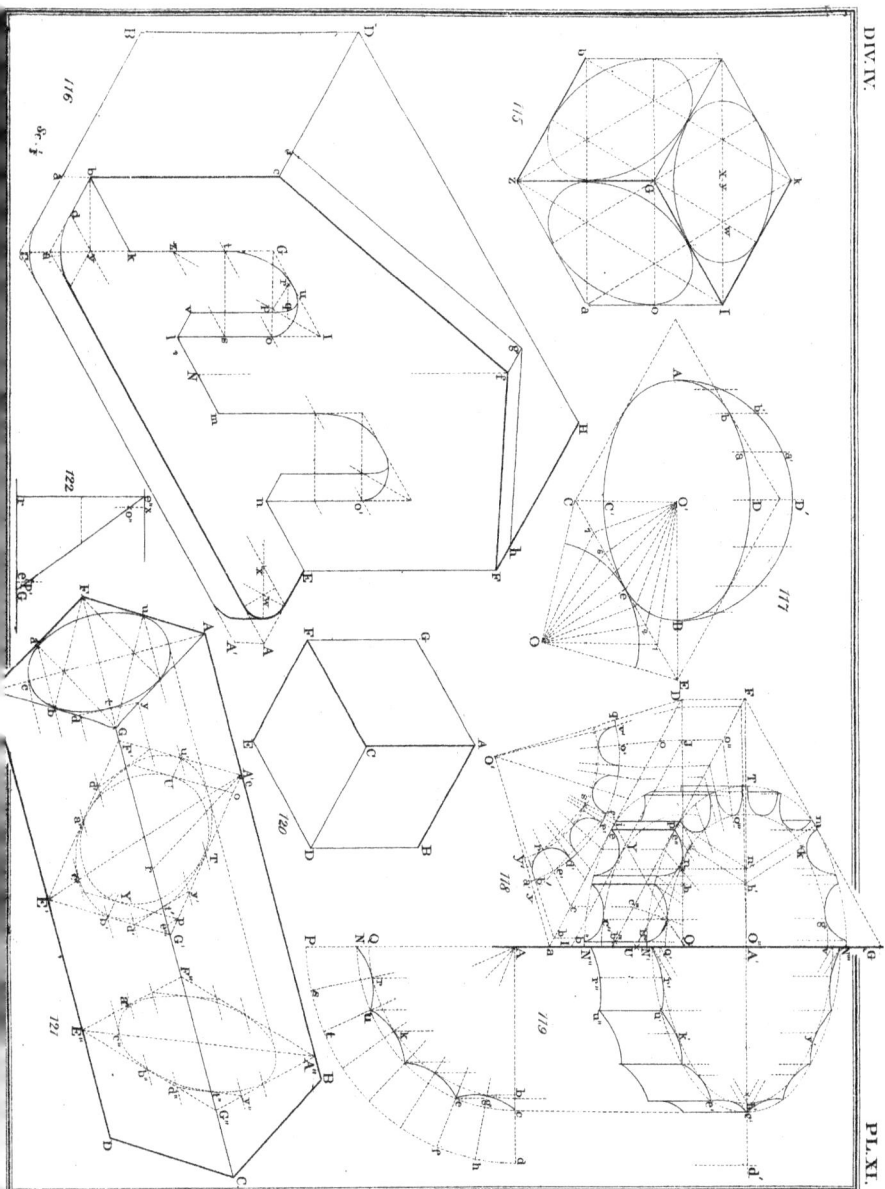

115

116

117

118

119

120

121

122

DIVISION FIFTH.

ELEMENTARY STRUCTURAL DRAWING.

<hr>

237. *Note.* The objects of this DIVISION are, to acquaint the student with a few things respecting the drawing of whole structures which are not met with in the drawing of mere details; to serve as a sort of review of practice in certain processes of execution; and to afford illustrations of parts of structures whose names have yet to be defined. Proceeding with the same order as regards material that was observed in DIVISION THIRD, we have:—

CHAPTER I.

STONE STRUCTURES.

238. EXAMPLE 1°. **A brick segmental Arch.** Pl. XII., Fig. 123. *Description of the structure.*—A segmental arch is one whose curved edges, as aCc, are less than semicircles. A brick segmental arch is usually built with the widths of the bricks placed radially, since, as the bricks are rectangular, the mortar is disposed between them in a wedge form in order that each brick with the mortar attached may act as a wedge; while if the length of the bricks be radial, the mortar spaces will be inconveniently wide at their outer ends, unless the arch be a very wide one, or unless it have a very large radius.

The *permanent* supports of the arch, as nPT, are called *abutments*, and the radial surface, as nab, against which the arch rests, is called a *skew-back.*

The temporary supports of an arch while it is being built are called *centres* or *centrings*, and vary from a mere curved frame made of pieces of board—as used in case of a small drain or round topped window—to a heavy and complicated framing, as used for the temporary support of heavy stone bridges.

Note. The general designing of these massive centrings may call for as much of scientific engineering *knowledge*, and their details and management may call for as much practical engineering skill, as does the construction of the permanent works to which these centrings are auxiliary. In short, the *detailed* design and management of auxiliary constructions, in general, is no unimportant department of engineering study.

The span is the distance, as *ac*, between the points of support, on the under surface of the arch. The stones over the arch and abutment, form the *spandril*, or *backing*, Q*d*P.

239. *Graphical construction.*—Let the scale be one of four feet to the inch=48 inches to one inch=$\frac{1}{48}$. Draw RT to represent the horizontal surface on which the arch rests. Let the radius of the inner curve of the arch be 7 feet, the height of the line *ac* from the ground 2 feet 8 inches, and the span 7 feet. Then at some point of the ground line, draw a vertical line, OC, for a centre line; then draw the abutments at equal distances on each side of the centre line, and 6 feet 8 inches apart. Let them be 2 feet 6 inches wide.

Since the span and radius have been made equal, O*b* and O*d* may be drawn, in this example, with the 60° triangle. Drawing these lines, and making O*a*=7 feet, make *ab* =one foot, draw the two curves at the end of the arch, and make *b* and *d* points in the top surfaces of the abutments.

To locate the bricks, since the thickness of the mortar between the bricks, at the inner curve of the arch, would be very slight, lay off two inches on the arc *aCc* an exact number of times. The distance taken in the compasses as two inches, may be so adapted as to be contained an exact number of times in *aCc*, since the thickness of the mortar has been neglected, but would in practice be so adjusted, as to allow an exact number of whole bricks in each course.

The arch being a foot thick, there will be three rows of bricks seen in its front. Draw therefore two arcs, dividing *ab* and *cd* into three spaces of four inches each, and repeat the process of division on both of them.

Having all the above-named divisions complete, fasten a fine needle vertically at O, and, keeping the edge of the ruler against it, to keep that edge on the centre without difficulty, draw the lines which represent the joints in each of the three courses of brick.

240. Ex. 2°. **A semi-cylindrical Culvert, having vertical**

quarter-cylindrical Wing Walls, truncated obliquely. Pl. XII., Fig. 124.

Description of the structure.—A culvert is an arched passage, often flat bottomed, constructed for the purpose of carrying water under a canal or other thoroughfare. Wing walls are curved continuations of the vertical flat wall in which the end of the arch is seen. Their use is to support the embankment through which the culvert is made to pass, and to prevent loose materials from the embankment from working their way or being washed into the culvert. Partly, perhaps, for appearance's sake, the slope of the plane which truncates the flat arch-wall, called the *spandril* wall, and the wing walls, is parallel to the slope of the embankment. The wing walls are often terminated by rectangular flat-topped posts—" piers" or " buttresses," AA', and the tops, both of these piers and of the walls, are covered with thin stones, $abcd$—$a''b''c''d''$, broader than the wall is thick, and collectively called the *coping*.

Since the parts of stone structures are not usually firmly bound or framed together, each course cannot be regarded as one solid piece, but rather each stone, in case, for instance, of the lowermost course, rests directly on the ground independently of other stones of the same course, hence if the ground were softer in some spots, under such a course, than in others, the stone resting on that spot would settle more than others, causing, in time, a general dislocation of the structure. Hence it is important to have what are called continuous bearings, that is, virtually, a single solid piece of some material on which several stones may rest, and placed between the lowest course and the ground.

Timbers buried away from the air are nearly imperishable; hence, timbers laid upon the ground, if that be firm, and covered with a double floor of plank, form a good foundation for stone structures; and in the case of a culvert, if such a flooring is made continuous over the whole space covered by the arch, it will prevent the flowing water from washing out the earth under the sides of the arch.

When the wing walls and spandril are built in courses of uniform thickness, the arrangement of the stones forming the arch, so as to bond neatly with those of the walls, offers some difficulties, as several things are to be harmonized. Thus, the arch stones must be of equal thickness, at least all except the top one, and then, there must be but little difference between the widths of the top, or *key stone*, and the other stones; the stones must not be disproportionately thin or very wide, they should have no re-entrant angles, or very acute angles, and there must not be any great extent of unbroken joint.

241. *Graphical construction.*—Let the scale be that of five feet to an inch = 60 inches to an inch = $\frac{1}{60}$.

a. Draw a centre line, BB', for the plan.

b. Supposing the radius of the outer surface, or back, of the arch to be $5\frac{5}{6}$ feet, draw CC' parallel to BB' and $5\frac{1}{2}$ feet from it.

c. Draw BE, and on C'C produced, make EC = 9 feet 8 inches, CD, the thickness of the face wall of the arch = 2 feet 4 inches, and the radius, *o*D, of the face of the wing wall = 4 feet.

d. With *o* as a centre, draw the quadrants CG, and DF, and with a radius of 3 feet 8 inches, draw the arc *ch*, the plan of the inner edge of the coping. Also draw at D and C, lines perpendicular to BB' to represent the face wall of the arch.

e. At G, draw G*h* towards *o*, and = 3 feet, for the length of the cap stone of the buttress, AA', and make its width = 2 feet 10 inches, tangent to CG at G. The top of this cap stone, being a flat quadrangular pyramid, draw diagonals through G and *h*, to represent its slanting edges.

f. Supposing the arch to be $1\frac{1}{2}$ feet thick, make C'H = $1\frac{1}{2}$ feet, and at C' and H, draw the irregular curved lines of the broken end of the arch, and the broken line near the centre line, also a fragment of the straight part of the coping.

g. Let the horizontal course on which the arch rests, be 2 feet 9 inches wide, i. e., make H*e* = 3 inches, and C'*n* = 1 foot; and let the planking project 3 inches beyond the said course, making *er* = 3 feet. Through *e*, *n* and *r*, draw lines parallel to BB' and extending a little to the right of C'H.

h. Proceeding to represent the parts of the arch substantially in the order of their distance from the eye, as seen in a plan view, a portion of the planking may next be represented. The pairs of broken edges, and the position of the joints, show that there are two layers of plank and that they break joints.

i. Under these planks, appear the foundation timbers, which being laid transversely, and being one foot wide and one foot apart, are represented by parallels one foot apart, and perpendicular to BB'. Let the planking project 4 inches beyond the left hand timber. Observe that two timbers touch each other under the arch front.

j. The general arrangement of stones in the curved courses of the wing wall, in order that they may break joints, is, to have three and four stones, respectively, in the consecutive courses. To indicate

124

123 - Sc. 1/48

125 - Sc. 1/48

this arrangement in the plan, hG, fb, gd and DC will represent the joints of alternate courses, and the lines km, &c. midway between the former, will represent the joints of the remaining intermediate courses.

This completes a partial and dissected plan which shows more of the construction than would a plan view of the finished culvert, and as much, as if the parts on both sides of the centre line were shown. In fact, in drawings which are strictly working drawings, each projection should show as much as possible in regard to each distinct part of the object represented.

242. *Passing to the side elevation*, which is a sectional one, showing parts in and beyond a vertical plane through the axis of the arch, we have :—

a. The foundation timbers, as $m'q$, &c., projected up from the plan ; or, one of them being so projected, the others may be constructed, independently of the plan, by the given measurements.

b. The double course of planking op, appears next with an occasional vertical joint, showing where a plank ends.

c. The buttress, A, and its cap stone Y, are projected up from the plan, and made 6 feet high, from the planking to G'.

d. From G' and h', the slanting top of the wing walls are shown, as having a slope of $1\frac{1}{2}$ to 1—i. e. $h'h'' = \frac{3}{2} h''u$—and the vertical lines at C', D' and D'' are projected up from C, D and D'''.

The remaining lines of the side elevation are best projected back from the end elevation, when that shall have been drawn.

243. *In the end or front elevation*, we have :—

a. At $m''m'''$, a side view of one of the foundation timbers, broken at m''', so as to show other timbers behind it.

b. The planking $o'o''$ in this view, shows the ends of the planks in both layers—breaking joints.

c. $No' = Bo'''$, taken from the plan ; and in general, all the horizontal distances on this elevation, are taken from the plan, on lines perpendicular to BB'.

d. The vertical sides of the buttress, A', are thus found. The heights of its parts are projected over from the side elevation.

e. The thickness of the foundation course, $ts = 1\frac{1}{2}$ feet, and $tr' = en$, on the plan.

f. The centre, O, of the face of the arch, is in the line $r't$ produced. The radius of the inner curve (intrados) of the arch is 4 feet and of the cylindrical back, behind the face wall, $5\frac{1}{2}$ feet—shown by a dotted arc. In representing the stones forming the arch, it is to be remembered that they must be equal, except the " key stone," *g*, which *may* be a little thicker than the others; they must also be of agreeable proportions, free from very acute angles, or from re-entrant obtuse angles; and must interfere as little as possible with the bond of the regular horizontal courses of the wing walls. There must also be an odd number of stones (ring stones) in the front of the arch.

On both elevations, draw the horizontal lines representing the wing wall courses as one foot in thickness, and divide the inner curve of the arch into 15 equal parts. Draw radial lines through the points of division. Their intersections with the horizontal lines are managed according to the principles just laid down.

g. The points, as *k* and *f*, in the plan, are then projected into the alternate courses of the side elevation, and into the line, Bo''', of the plan.

From the latter line, the several distances, $o'''b''$, &c., from o''', thus found, are transferred to the line o'N, as at $o'b''''$, &c., and at these points the vertical joints of the front elevation are drawn in their proper position, as being the same actual joints, shown by the vertical lines of the side elevation. In the stones immediately under the coping, there must generally be some irregularity, in order to avoid triangular stones, or stones of inappropriate size.

h. To construct the front elevation of the coping. All points, as *a*, a', a'', in either the front or back, or upper or lower edges of the coping, are found in the same way, and as follows:

a'' is in a horizontal line through a' and in a line $a''a''''$, whose distance from o' equals the distance $o'''a'''$ on the plan. Construct-ing other points similarly, the edges of the coping may be drawn with an "irregular curve."

The horizontal portion of the coping, over the arch, is projected over from C$'$ and from the two ends of the vertical line at D$''$.

Execution.—In respect to this, the drawing explains itself.

CHAPTER II.

244. Ex. 3°. Elevation of a "King Post Truss."

Mechanical construction, &c.—A Truss is an assemblage of pieces so fastened together as to be virtually a single piece, and therefore exerting only a vertical force, due to its weight, upon the supporting walls.

In Pl. XII., Fig. 125, A is a *tie beam;* B is a *principal;* C is a *rafter;* D is the *king post;* E is a *strut;* F is a *wall plate;* G is a *purlin*—running parallel to the ridge of the roof, from truss to truss, and supporting the rafters. H is the *ridge pole;* W is the wall, and *ab* is a strap by which the tie beam is suspended from the king post.

245. *Graphical construction.*—In the figure, only half of the truss is shown, but the directions apply to the drawing of the whole. In these directions an accent, thus ′ , indicates feet, and two accents, ″ , inches.

a. Draw the vertical centre line *b*D.

b. Draw the upper and lower edges of the tie beam, one foot apart, and 12′ in length, on each side of the vertical line.

c. On the centre line, lay off from the top of the tie beam, 5′—6″ to locate the intersection of the tops of the principals; and on the top of the tie beam, lay off 11′ on each side, to locate the intersection of the upper faces of the principals with the top of the tie beam.

d. Draw the line joining the two points just found, and on any perpendicular to it, as *fg*, lay off its depth = 8″, and draw its lower edge parallel to the upper edge. Make the shoulder at *o* = 3″ and parallel to *fg*.

e. From the top of the beam, draw short indefinite lines, *c*, 6″ each side of the centre line, and note the points; as *e*, where they would meet the upper sides of the principals.

f. Draw vertical lines on each side of the centre line and 4″ from it.

g. From the points, as *e*, draw lines parallel to *fg* till they intersect the last named vertical lines.

h. Make $ns=5'—9''$. Make the short vertical distance at $c=4''$, draw *sc*, and make the upper side of the strut parallel to *sc*, and $4''$ from it. Note the intersection of this parallel with the line to the left of D, and connect this point with the upper end of *c*, to complete the strut.

i. Draw the edges of the rafter, parallel to those of the principal, $4''$ apart, and leaving $4''$ between the rafter and the principal. At *o*, draw a vertical line till it meets the lower edge of C, and from this intersection draw a horizontal line till it meets the upper edge of C; which gives proper dimensions to the wall plate.

j. From the intersections of the upper edges of the rafters, lay off downwards on the centre line $12''$, and make the ridge pole, thus located, $3''$ wide.

k. In the middle of the upper edge of the principal, place the purlin $4''\times6''$, and setting $2''$ into the principal.

l. Let the strap, *ab*, be $2''$ wide, and $2'—6''$ long from the bottom of the tie beam. Let it be spiked to the king post and tie beam, and let it be half an inch thick, as shown below the beam. W, the supporting wall, is made at pleasure.

Execution.—This mainly explains itself. As working drawings usually have the dimensions figured upon them, let the dimensions be recorded in small hair line figures, between arrow heads which denote what points the measurements refer to.

246. Ex. 4°. **A "Queen Post Truss" Bridge.** Pl. XIII. Fig. 126.

Mechanical construction.—This is a bridge of 33 feet span, over a canal $20'—6''$ wide between its banks at top, and $20'—2''$ at the water line. It rests on stone abutments, R and P, one of which is represented as resting on a plank and timber foundation, the other on " piles."

A is the tie beam; B, B' the *queen posts ;* C, C' the principals; D the *collar beam,* or *straining sill ;* R, P, the abutments; *eQt* the pavement of the *tow path ; t*K the stone side walls of the canal; TT the opposite timber wall, held by timbers UU', N, dovetailed into the wall timbers; E, S, the piles, iron shod at bottom. These are the principal parts.

247. *Graphical construction.*—Let the scale be one of five feet to the inch.

a. All parts of the truss are laid off on, or from, the centre line AD. A is $14''$ deep; the dimensions of BB' are $12''\times6'$, except at top, where they are $10''$ wide for a vertical space of $16''$. C and D

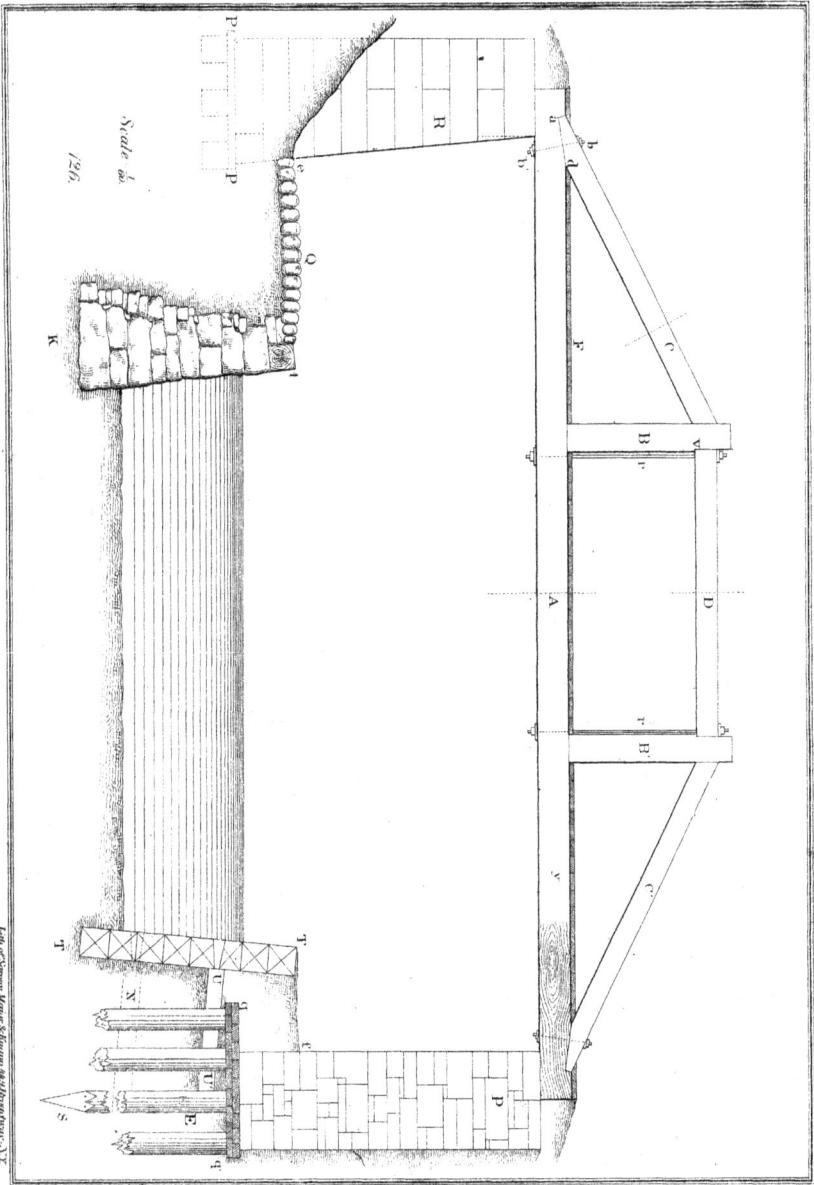

Scale ⅟₂₆

Lith. of Sarony Major & Knapp, 449 Broadway, N.Y.

are each 10″ deep. BB′ are 10′ apart, and the feet of C and C′, 12″ from the ends of the tie beam, which is 36′ long. D is 6″ below the top of the queen posts. *rr* are inch rods with five inch washers, ¾″ thick, and nuts 2½″ × 1″. *bb′* is a ¾″ bolt; with washer 4″ × ¾″ and nuts, 2″ × 1″; and perpendicular to the joint, *ad*.

b. From each end of the tie beam, lay off 1′—9″ each way for the width of the abutments, at the top. Make the right hand abutment rectangular in section and 11′ high, of rectangular stones in irregular bond (76). Let the left hand abutment have a batter of 1″ in 1′ on the side towards the canal, and let it be eleven feet high, in eleven equal courses.

c. Make *et*, the width of the paved tow path = 7′—6″, with a rise in the centre, at Q, of 6″.

d. The side wall is of rubble, 4′ thick at bottom, and extending 18″ below the water, with a batter of 1″ in 1′, and having its upper edge formed of a timber 12″ square.

e. The right hand abutment rests on a double course of three-inch planks, *qq′*, 5½′ broad, and resting on four rows of 10″ piles, ES. S is the sheet iron conical shoe at the lower end of one of these piles, the dots at the upper end of which represent nails which fasten it to the pile.

f. TT is a timber wall having a batter of 1″ to 1′, and held in place by timbers, UU′, N, dovetailed into it at its horizontal joints, in various places.

g. The water line is 2′ below T*t*, and the water is 4½ feet deep.

248. *Execution.*—It is intended that this plate should be tinted, though, on account of the difficulty of procuring adequate engraved fac-similes of tinted hand-made drawings, it is here shown only as a finished line drawing, and as such, explains itself, after observing that as the left hand abutment is shown in elevation, it is dotted below the ground ; while, as the right hand abutment is shown in section, it is made wholly in full lines, and earth is shown only at each side of it.

The usual conventional rule is, to fill the sectional elevation of a stone wall with wavy lines ; but where other marks serve to distinguish elevations from sections, as in the case just described, this labor is unnecessary.

The following would be the general order of operations, in case this drawing were shaded.

a. Pencil all parts in fine faint lines.

b. Ink all parts in fine lines.

c. Grain the wood work with a very fine pen and light indian ink, the sides of timbers as seen on a newly-planed board, the ends of large timbers in rings and radial cracks, and the ends of planks in diagonal straight lines. See also the figures at y, where the lines of graining outside of the knots, are to extend throughout the tie beam.

d. Tint the wood work—the sides with pale clear burnt sienna, the ends with a darker tint of burnt sienna and indian ink.

e. Tint the abutments, and other stone work, with prussian blue mixed with a little carmine and indian ink, put on in a very light tint.

f. Grain the abutments in waving rows of fine, pale, vertical lines of uniform thickness, about one sixteenth of an inch long, leaving the upper and left hand edges of the stones blank, to represent the mortar. The part of the left hand abutment which is under ground is dotted only, as in the plate.

g. Grain the canal walls and paving, as shown in the plate, to indicate boulder rubble.

h. Shade the piles roughly, they being roughly cylindrical; tint them with pale burnt sienna, and the shoe, S, with prussian blue, the conventional tint for iron.

i. Rule the water in blue lines, distributed as in the figure.

j. Tint the dirt in fine horizontal strokes of any dingy mixture, in which burnt sienna prevails, in the parts above the water, and ink, in the muddy parts below the water, and then add, or not, the pen strokes shown in the plate, to represent sand, gravel, &c.

k. Place heavy lines on the right hand and lower edges of all surfaces, except where such lines form dividing lines between two surfaces in the same plane. A heavy line on the under side of the floor planks, indicates that those planks project beyond the tie beam A.

249. Ex. 5°. **Construction from a Model.** Pl. XIV., Figs. 127, 128. *General Description.*—This plate contains two elevations of an architectural Model. It is introduced as affording excellent practice in tinting and shading large surfaces, and useful elementary studies of shadows. The construction of these elevations from given measurements is so simple, that only the base and several centre lines need be pointed out.

QR is the ground line. ST is a centre line for the flat topped tower in Fig. 127. UV is a centre line for the whole of Fig. 128, except the left hand tower and its pedestal. WX is a centre line for the tower through which it passes. YZ is the centre line for

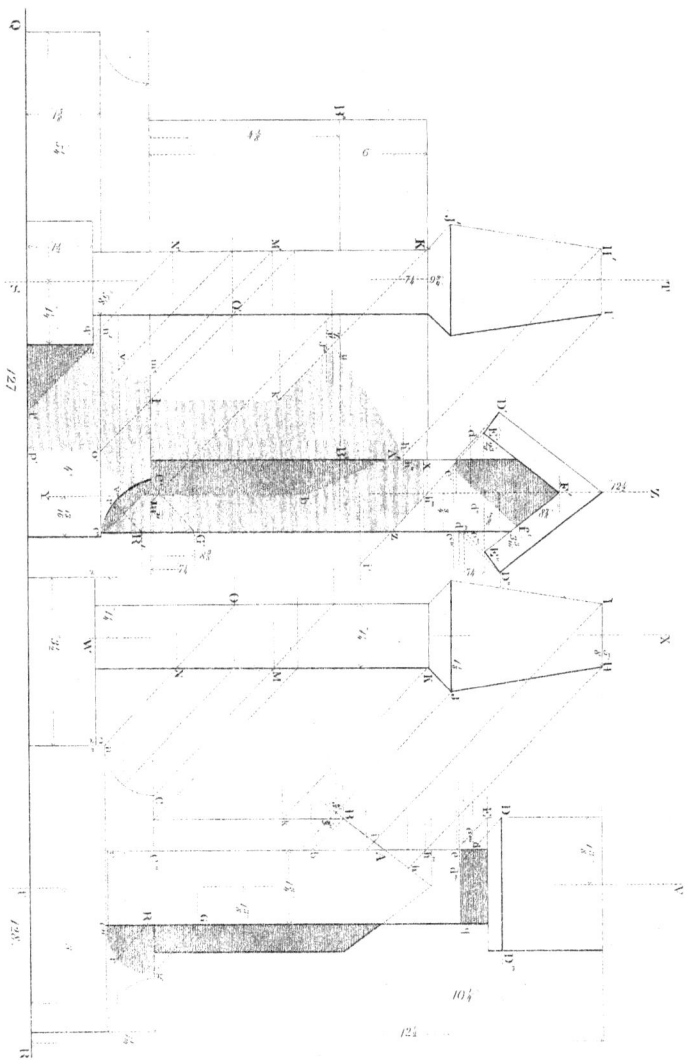

the roofed tower in Fig. 127. The measurements are recorded in full, referred to the centre lines, base line, and bases of the towers, which are the parts to be first drawn.

250. *Graphical construction of the shadows.*

1°. The roof, D—D'D'', casts a shadow on its tower. The point, EE', casts a shadow where the ray, E*e*, pierces the side of the tower. *e* is one projection of this point; *e'*, the other projection of the same point, is at the intersection of the line *ee''e'* with the other projection, E'*e'*, of the ray. The shadow of a line on a parallel plane (162) is parallel to itself, hence *e'f'*, parallel to E'F', is the shadow of E'F'.

The shadow of DE—D'E' joins *e'* with the shadow of D—D'. The point *d*, determined by the ray D*d*, is one projection of the latter shadow; the other projection, *d'*, is at the intersection of *dd''d'* with the other projection, D'*d'*, of the ray. *d'* is on the side of the tower, produced, hence *e'd'* is only a real shadow line from *e'* till it intersects the edge of the tower.

Remembering that the direction of the light is supposed to change with each position of the observer, so that as he faces each side of the model, in succession, the light comes from left to right and from behind his left shoulder, it appears that the point, DD'', casts a shadow on the face of the tower, seen in Fig. 128, and that D''*d''''* will be the position of the ray, through this point, on Fig. 127. The point *d'''* is therefore one projection of the shadow of DD''. The other is at *d''''*, the intersection of the lines *d'''d''''* with D*d''''*, the other projection of the ray. Likewise EE'' casts a shadow, *e'''e''''*, on the same face of the tower, produced. DD''', being parallel to the face of the tower now being considered, its shadow, *d''''q*, is parallel to it. The line from *d''''* towards *e''''*, till the edge of the tower, is the real portion of the shadow of DE—D''E''.

251. From the foregoing it will be seen how most of these shadows are found, so that each step in the process of finding similar shadows will not be repeated.

2°. The body of the building—or model—casts a shadow on the roofed tower, beginning at AA' (160). The shadow of BB' on the side of this tower is *bb'*, found as in previous cases, and A'*b'* is the shadow of AB—A'B'. From *b'* downwards, a vertical line is the shadow of the vertical corner edge of the body of the model upon the parallel face of the tower.

3°. The line CC''—C', which is perpendicular to the side of the roofed tower, casts a shadow, C'*c'*, in the direction of the projection of a ray of light on the side of the tower.

4°. In Fig. 127, a similar shadow, $s't'$, is cast by the edge s'—ss'' of the smaller pedestal.

5°. In Fig. 128, is visible the curved shadow, $c''rg$, cast by the vertical edge, at c'', of the tower, on the curved part of the pedestal of the tower. The point g is found by drawing a ray, $G'C'$—Gg, which meets the upper edge of the pedestal at gC'. The point c'', the intersection of the edge of the tower with the curved part of the pedestal, is another point. Any intermediate point, as r, is found by drawing the ray $R'r'$; r' is then one projection of the shadow of $R'R$, and the other is at the intersection of the line $r'r$ with the other projection Rr of the ray. These are all the shadows which are very near to the objects casting them.

6°. The flat topped tower casts a shadow on the roof of the body. The upper back corner, HH', casts a shadow on the roof, of which h is one projection and h' the other. The back upper edge H—$H'I'$ being parallel to the roof, the short shadow $h'h''$, leaving the roof at h'', is parallel to $H'I'$. The left hand back edge HJ—$H'J'$ casts a shadow on the roof, of which hh' is one point. The point JJ' casts a shadow jj' on the roof produced. $h'j'$ is therefore a real shadow only till it leaves the actual roof at u.

7°. The same tower casts a shadow on the vertical side of the body, of which $j''j'''$, found as in previous cases, and u, are points. The upper back point, KK', of the shaft of the tower, casts a shadow, kk', which is joined with j''', giving the shadow of JK—$J'K'$. From k', $k'l$ is the vertical shadow line of the left hand back edge of the flat topped tower on the parallel plane of the side of the body of the model.

8°. The same tower casts a shadow on the curved—cylindrical—part of the pedestal. To find the point m', of shadow, draw a ray, MC, Fig. 128, intersecting the upper edge of the pedestal at C, which is therefore one projection of the shadow of the point MM'. The other projection, m', of the same shadow, is at the intersection of the other projection, $M'm'$, of the ray, with the other projection, $m'C'$, of the edge of the pedestal. The point of shadow, nn', cast by the point NN', is similarly found, and so is the point oo', cast by the point OO' of the front right hand edge of the tower. Make $m'm''=n'o'$, and find intermediate points, $v'v''$, as rr' was found, and the curved shadow on the cylindrical part of the pedestal will then be found.

9°. From n' and o', vertical lines are the shadows of opposite diagonal edges of the tower, on the vertical face of the main pedestal.

10⁰. This flat topped tower also casts a shadow on the side of the roofed tower. The right back corner, H—I′, of the top, casts the shadow $h'''h''''$ on the side of the roofed tower, through which the shadow line, $h''''x$ is drawn, parallel to the line H—H′I′ which casts it. The right hand top line, I′—IH, being perpendicular to the plane of the sides of this tower, casts the shadow $h''''i'$ upon it, parallel to the projection of a ray of light. (162.) This shadow line is real, only till it leaves the tower at z ;—i' being in the plane of the side of the tower produced—and it completes all the shadows visible in the two elevations.

7

CHAPTER III.

252. Ex. 6⁰. A Railway Track. Pl. XV., Figs. 129–134.

Mechanical construction, &c.—It may be thought an oversight to style this plate the drawing of a railroad track; but taking the track alone, or separate from its various special supports, as bridges, &c., its graphical representation is mainly summed up in that of two parts; *first*, the union of two rails at their joints; *second*, the intersection of two rails at the crossing of tracks, or at turn-outs. The fixture shown in Fig. 129, placed at the intersection of two rails to allow the unobstructed passage of car wheels, in either direction on either rail, is called a "Frog." Let y and z be fragments of two rails of the same track, then the side Hf of the point of the frog, and the portion $k\,k'$ of its side flange, B, are in a line with the edges, denoted by dots, of the rails y and z, so that as the wheel passes either way, its flange rolls through the groove, I, without obstruction. When the wheel passes from y towards z there is a possibility of the flange's being caught in the groove, J, by dodging the point, f. To guard against this, a guard rail, $g\,g$, is placed near to the inside of the other rail, supposed to be on the side of the frog towards Fig. 132, as shown in the small sketch, Fig. 132, which prevents the pair of wheels, or the car-truck, from working so far towards the flange, B, as to allow the flange of the wheel to run into the groove, J, and so run off the track. F f, and the portion, $l\,l'$, of the flange, A, are in a line with the inner edge of the rail of a turn-out, for instance, the opposite rail being on the side of the frog towards the upper border of the plate, as shown in Fig. 132. Hence the flange of a car wheel in passing in either direction on the turn-out, passes through the groove, J, and is prevented from running into the groove, I, by a guard rail, near the inner edge of the opposite turn-out rail, as at U, Fig. 132.

253. Fig. 130 represents the under side of the right hand portion of the frog, and shows the nuts which secure one of the bolts which secure the steel plates, as D, E; bolts whose heads, as at u and v, are smooth and sunk into the plates so that their upper surfaces are

flush. It will be seen that there are two nuts on each bolt, as at D', on the bolt u—D D', which appears below the elevation, since it occurs between two of the cross-ties (sleepers) of the track. The nuts, as L, belonging to the bolt, b'', which are in the chairs, $q'p'$, w', x', are sunk in cylindrical recesses in the bottom of the frog, so as not to interfere with the cross-tie on which the surface, L, rests. The extra nut is called a check or "jam" nut. When screwed on snugly it wedges the first nut and itself also against the threads of the screw, so that the violent tremulous motion to which the frog is subjected during the rapid passage of heavy trains cannot start either of them.

In the end elevation, Fig. 131, A is the recess in the chair $x\ x'$, fitted for the reception of the rail, and B is the end of a rail in its place, as shown at y in the plan.

254. *Graphical Construction.*—From the above description it follows that the whole length of the frog depends on the shape of the part H f F, and the distance between this part and the side rails, as $c\ l$. In the present example $a\ c=1'$—$11''$ and $cf = 20''$. $e\ d$ is $11''$ and $n\ k$ is $2''$ from F f. Having these relations given, and knowing that the lines at the extreme ends are perpendicular to the rails at those ends, the several figures of the frog can be constructed from the given measurements, without further explanation.

255. Fig. 133, is an isometrical drawing—scale $\frac{1}{12}$—of a recent somewhat elaborate and very secure mode of connexion of the common solid or H rails. A section of the Boston and Worcester R.R. recently (Nov. 1857) laid in this manner, allows the cars to ride over it, now, 1861, with great smoothness of motion and freedom from the loud clack which accompanies the use of ordinary chairs. A, A, A, are the sleepers (cross-ties), D is a stout oak plank, perhaps six feet long, resting on three sleepers, and fitted to the curved side of the rail, as shown at d. This plank is on the outside of the track. On the inner side the rails are spiked in the usual way with hook-headed spikes $s\ s\ s$, of which those at the joint, r, pass through a flat wrought iron plate, P, which gives a better bearing to the end of the rail, and prevents dislocation of parts. Each plank, as D, is bolted to the rail by four horizontal half inch bolts, b, b, b, b, furnished with nuts and washers on the further side of D (not seen).

Note.—A recent (1860) modification of the above construction consists in substituting for the plate P, a short piece or strap of iron fitted to the surface of the inside of the rail, and through which the two bolts bb, next to the joint, pass.

Sometimes, indeed, such a short strap is placed on each side of the rail, and then the beam D is dispensed with. But wood is deemed preferable, since its elasticity renders the bolts, b, less likely to work loose than in a rigid iron construction.

In connection with a very solidly constructed section of the above railroad, on the plan just described, the experiment was made of having the track break joints. That is, a joint, as a, Fig. 132, on one rail of a track, is placed opposite the centre of the rail bc of the other line of the same track.

As a track always tends to settle at the joints, a jumping motion is induced in a passing train, which perhaps may be thought to be less violent if only on one rail at a time. A reduced jumping on the rail would in turn diminish the tendency to settle at the joints.

256. *Graphical Construction.*—Three lines through X, making angles of 60° with each other, will be the isometric axes. Remembering that it is the *relative* position of the lines which distinguishes an isometrical drawing, we can place XX' parallel to the lower border, and thus fill out the plate to better advantage. The rail being 4″ wide at bottom, and 4″ high, circumscribe it by a square X*can*, from the sides of which, or from its vertical centre line, lay off, on isometric lines, the distances to the various points on the rail. Thus, let the widest part of the rail, near the top, be 3″ across, and $\frac{1}{2}$ an inch below the top ac. Let the width at the top be 2″, and at the narrowest part 1″; and let the mean thickness of the lower flange be $\frac{3}{4}$″. The sides of the rail are represented by the bottom lines at XX', and the tangents each side of R, to the curves of the section. Let the plank D be 6″ wide, and 4″ high. All the lines of the spikes, ss, are isometrical lines except their top edges, as st. The curve at the joint r, and at X', are similar to the corresponding parts of the section at X.

To secure ease of graphical construction, let the bolt heads, b, &c., be placed so that their edges shall be isometric lines.

257. Fig. 134, is a plan and end elevation of a heavy cast-iron chair lately (Nov. 1857) introduced upon the Troy and Boston R.R. On the outside of the rail, the top, ab, of the chair is flush with the top, bc, of the rail, and thus forms essentially a continuous rail, to do which, while retaining the simplicity of the solid rail, is a thing to be desired.

258. Ex. 7°. **The Hydraulic Ram.** In order to give an iron construction, from the department of machinery, so as to render this volume a more fit elementary course for the machinist as well

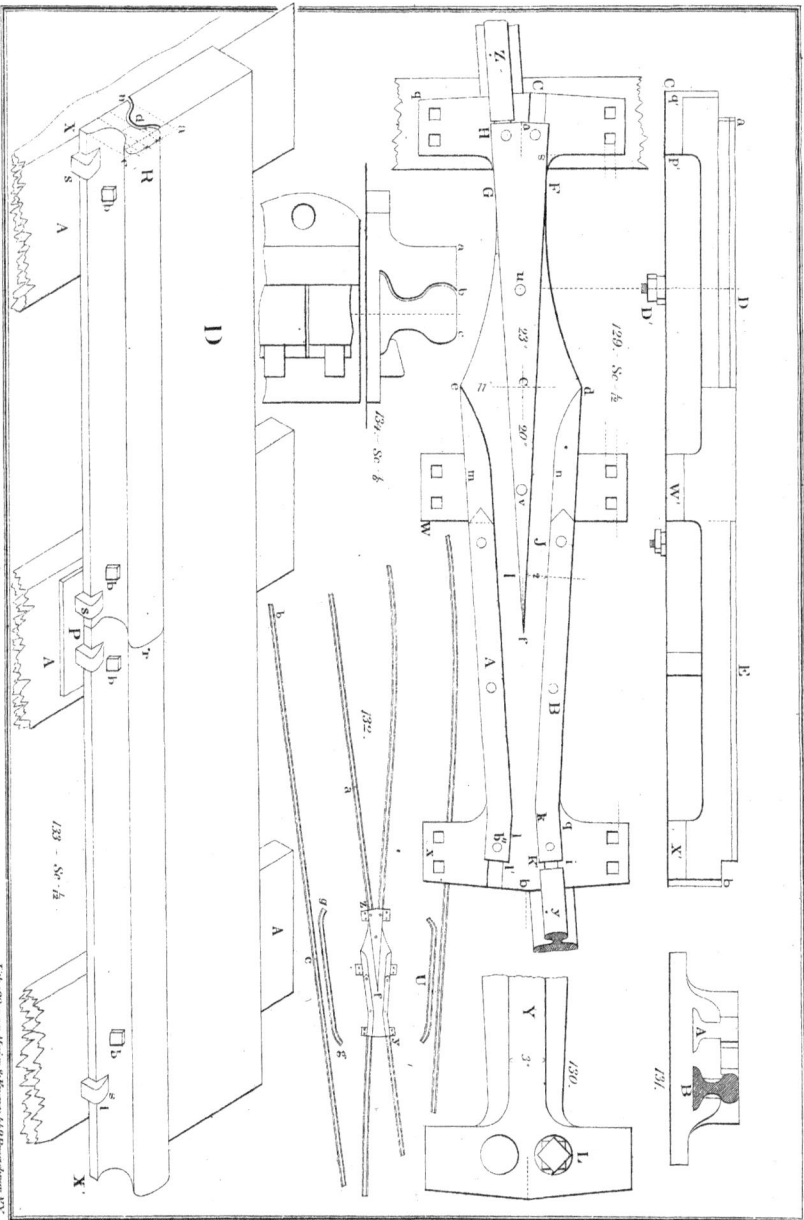

as for the civil engineer, a simple and generally useful structure, viz. a hydraulic ram, has been chosen, as a fit example with which to close the present volume.

This machine is designed to employ the power of running water to elevate water to any desired height.

Pl. XVI., Figs. 135–137, shows a hydraulic ram, of highly approved construction, and of half the full size.

259. *Mechanical construction.*—FF—F'F' are feet to support the machine. These are screwed to a floor or other firm support. AB—A'B'B' is the inlet pipe, opening into the air chamber C, at a—$a'b'$ and ending at dd—$d'd'$—$d''d''$ the opening in the top of the waste valve chamber, E—E'—E''. At a—$a'b'$ is the opening as just noticed from the inlet pipe into the air chamber C (not seen in the plan). This opening is controlled by a leather valve ee', weighted with a bit of copper $e''e'''$, and is fastened by a screw $h''h'''$, and an oblong washer $g'g$. At N and H are the extremities of two outlet pipes leading from the air chamber at F''F'''. Either one, but not both of these outlet pipes together, may be used, as one of the exchangeable flanges, H' is solid, while the other is perforated, as seen at M', Fig. 137. The air chamber is secured by bolts passing through its flange $f'f'$, through the pasteboard or leather packing, pp—p', and the flange D—D'D' at cc. This flange, and part of the inlet pipe are shown as broken in the elevation, so as to expose the valve ee', and the adjacent parts. LL' is a flange through which the inlet pipe passes, and this pipe is slit and bent over the inner edge of the aperture in LL', forming a flange, which presses against a leather packing, tt', and makes a tight joint. The outlet pipes are secured in the same way. At uu—u' are the square heads of bolts which fasten the flanges to the projections UU—U'. K—K' is a shelf bearing the waste valve chamber, E—E'E'', and the adjacent parts. W—W' is the flange of this valve chamber, secured by two bolts at vv''—v', which pass through the leather packing y. $h'h''$ is the waste valve, perforated with holes, x, to allow water to flow through it. mm' is the valve stem. $d'd'k'k'$ is a perforated standard serving as a guide to the valve stem, and also as a support to the hollow screw s. n is a rest, secured to the valve stem by a pin p''. q'' is a nut, part of which, qq', is made hexagonal. r is a "jam" nut (253).

In the plan of this portion of the machine, the innermost circle is the top of the valve stem; next is the body of the valve stem; next, the top of the rest; next, the bottom of the same; next, the nut q''; and outside of that, and resting on the top of the waste valve chamber, are the standards, dd.

260. *Operation.—Principles involved.*—In the case of what might be called *passive constructions*, that is mere stationary supports, like bridges, &c., a knowledge of the construction of the parts enables one to proceed intelligently in making a drawing; but, in the case of what may, in opposition to the foregoing, be called *active constructions*, or machines, a knowledge of their mode of operation is usually essential to the most expeditious and accurate graphical construction of them, because a machine consists of a train of connected pieces, so that a given position of any piece implies a corresponding position for every other part. Having, then, in a drawing, assumed a definite position for some important part, the remaining parts must be *located* from a knowledge of the machine, though *drawn* by measurements of the dimensions of that part. Only *fixed bearings*, and *centres of motion*, can properly be located by measurement, in machine drawing.

The principles involved in the operation of the hydraulic ram may be summed up under three heads, as follows:

261. I. *Work.* a. When a certain *weight* is moved through a certain *space*, a certain amount of *work* is expended.

b. Thus; when a quantity of water descends through a certain space, a certain amount of work is developed.

c. As the idea of work involves the idea both of weight moved, and space traversed, it follows that *works* may be equal, while the weights and spaces may be unequal. Thus the work developed by a certain quantity of water, while descending through a certain height, may be equal to that expended in raising a portion of that water to a greater height.

262. II. *Equilibrium.* a. Where forces are balanced, or mutually neutralized, they are said to be in equilibrium. Now the usual fact is, that when such equilibrium is disturbed, it does not restore itself at once, but gradually, by a series of alternations about the state of equilibrium. Thus a stationary pendulum, being swung from its position of equilibrium, does not, at the first returning vibration, stop at the lowest point, but does so only after many vibrations.

b. Theoretically, these vibrations, as in the case of the pendulum, would never stop, but in practice the resistance of the air, friction, &c., make a continual supply of a greater or less amount of force necessary to perpetuate the alternations about the position or state of equilibrium.

263. III. A *physical fact* taken account of in the hydraulic ram, is, that water in contact with compressed air will absorb a certain portion of such air.

264. Passing now more particularly to a description of the operation of the hydraulic ram: 1°. Water from some elevated pond or reservoir flows into the machine, through the inlet pipe AA′, and continues through the machine, and flows out through the holes in the waste valve h′h″, pressing meanwhile against the solid parts of the roof of this valve, whose hollow form — open at the bottom — is clearly shown in Fig. 136.

2°. Presently the water acquires such a velocity as to press so strongly against the roof of the waste valve, that this valve is lifted against the under side of the roof of its chamber which it fits accurately.

3°. The water thus instantly checked, expends its acquired force in rushing through the valve e—e′e‴ and in compressing the air in the air chamber C.

4°. The holes F″ or F‴ of the outlet pipe, leading to an unobstructed outlet, the compressed air immediately forces the water out through the outlet pipe until, after a number of repetitions of this chain of operations, the portion of the water thus expelled from the air chamber is raised to a considerable height.

5°. In accordance with the second principle, the flow of water from the air chamber does not cease at the moment when the confined air is restored to its natural density, but continues, so that—taking account also of the absorption of the air by the water at the time of compression—for a moment the air of the air chamber is more rare than the external atmosphere. Hence to keep a constant supply of air to the air chamber, a fine hole called a snifting hole, is punctured, as with a needle, at ss′, i.e., just at the entrance of the inlet pipe into the machine. Through this hole air enters, with a snifting sound, when the flow of water recommences, so as to supply the air chamber with a constant quantity of air. When the waste valve is at the bottom of the chamber EE′, the nut and "jam" are together at the bottom of the screw s′, and the valve is at liberty to make a full stroke. By raising the valve to its highest point and turning the nut and "jam" to some position as shown in the figure, the stroke of the valve can be shortened at pleasure, and, at its lowest point, will be as far from the bottom of the chamber as the "jam," q″, is above its lowest position.

266. In practice, it is found that the strokes of the waste valve shortly become regular; their frequency depending in any given case on the height of the supply reservoir, the height of the ejected column, the size of the machine, the length of the stroke of the valve, &c.

267. The proportion of water discharged into the receiving reservoir will also depend on the above named circumstances, being more or less than one third of the quantity entering the machine at AA'. In a machine by M. Montgolfier of France, said to be the original inventor—water falling $7\frac{1}{2}$ feet, raised $\frac{2}{21}$ of itself to a height of 50 feet.

268. *Graphical Construction.*—Scale; half the full size. *a.* Having the extreme dimensions of the plan, in round numbers 9" and 12", proceed to arrange the ground line, leaving room for the plan below it.

b. Draw a centre line, NC, for plan and elevation, about in the middle of the width of the plate.

c. Draw a centre line, AK, for the plan, parallel to the ground line.

d. Exactly $4\frac{1}{2}$" from the centre line NC, draw the centre line vv" —K'm' for the waste valve chamber and parts adjacent.

e. With the intersection, *, of the centre lines of the plan, as a centre, draw circles having radii of $1\frac{3}{4}$" and $3\frac{1}{16}$" respectively, and through the same centre, draw diagonals, as cc.

f. On the centre line, NC, are the centres of the circles, F"F''', whose circumferences come within $\frac{1}{10}$ of an inch of the inner one of the two circles just drawn.

g. Draw the valve, e, the copper weight e", the screw end, h, and the nut and oblong washer, h" and g.

h. Locate, at once, the centres of all the small circles, cc, &c., by the intersection of arcs, $\frac{1}{4}$" from the circle pp having * for a centre, with the diagonals; then proceed to draw these circles.

i. Draw the projections, as U, drawing the opposite ones simultaneously, and using an auxiliary end view of the nuts u, as often explained before.

j. Draw the feet, F, with their grooves, F, and bevel edged screw holes, L".

k. In drawing the shelf, K, and flange W, the intersection of the centre lines BK and $m'm$, is the centre for the curves which intersect the centre line AK; the corners, 1 1, of the nuts, v,v", are the centres for the curves that cross the centre line, vv"; and the remaining outlines of the shelf are tangents to the arcs thus drawn, and those of the flange are lines sketched in so as to give curves tangent to the arcs already drawn, and short straight lines parallel to vv".

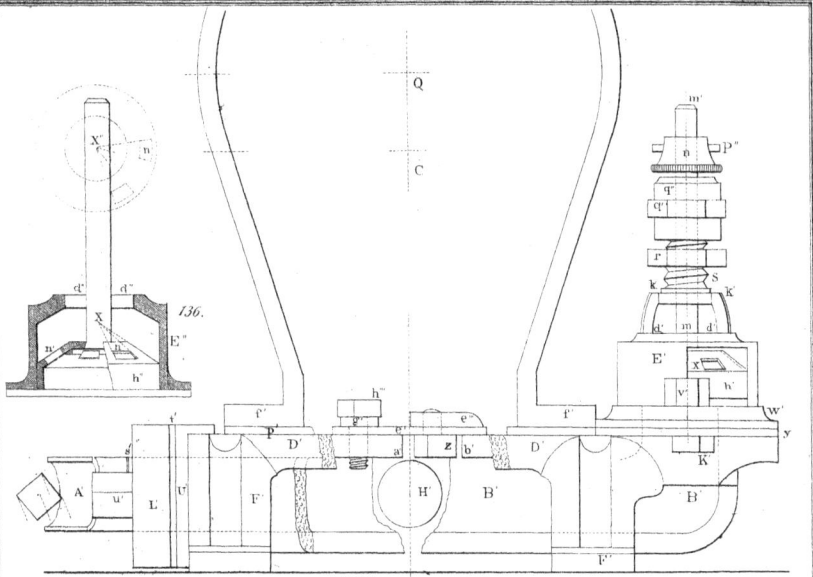

135.

136.

Sc. f

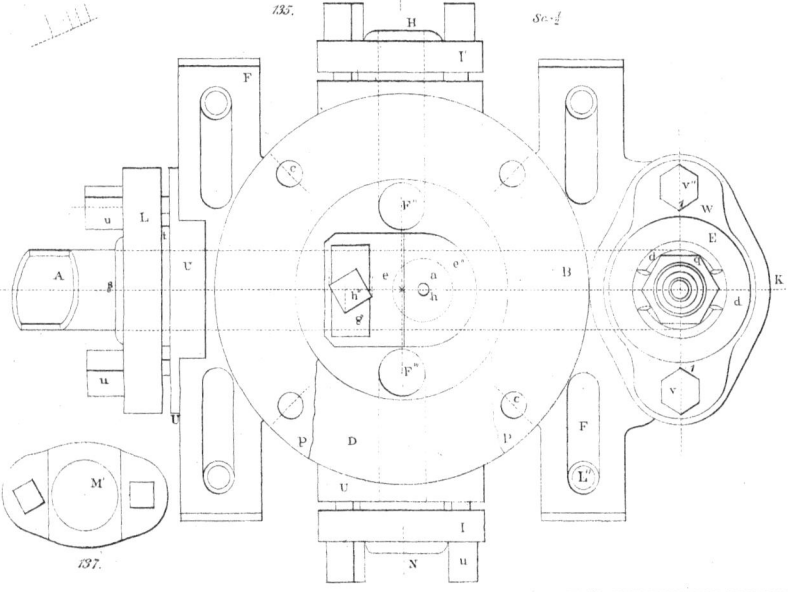

137.

l. The remaining circles and larger hexagon, *u'*, of this portion of the plan, have the intersection of the centre lines for a centre ; and may be drawn by measurements independently of the elevation, or by projection from the elevation, after that shall have been finished.

269. Passing to the elevation :—

a. Construct, at one position of the T square, the horizontal lines of both feet ; then the horizontal lines of the nuts *u'*, and flange L', and projection U' ; with the horizontal lines of the floor of the air chamber and adjacent parts.

b. Project up from the plan the vertical edges of the feet, F'F', the flange, nut, and projection L', *u'* and U', the valve *e'*, the copper *e''*, the screw *h''*, the washer *g*, the air chamber flange *f'f'*, and screw z. Break away the portion D—see plan—of the body of the machine, and the near wall of the water channel A'B'. Break away also the further wall of the water channel so as to show a section, H', of the further outlet pipe, H—see plan. Q is the centre of the spherical part of the air chamber to which the conical part is tangent.

c. Draw all the horizontal lines of the waste valve chamber and parts adjacent. Make the edges of the threads of the screw straight and slightly inclined upwards toward the right.

d. Project up from the plan, or lay off, by measurement, the widths of various parts through which the valve stem passes, and draw their vertical edges.

Fig. 136 is a section of the waste valve chamber, showing part both of the interior and exterior of the waste valve. The dotted circles form an auxiliary plan of this valve, in which the holes have two radial sides, and two circular sides with *x''* as a centre. The top of the valve is conical, so that in the detail below, two of the sides of the hole *n*, tend towards the vertex, *x*. At *n'*, one of these holes, of which there are supposed to be five, is shown in section.

Fig. 137. The outlines of M, one of the outlet pipe flanges, are drawn by processes similar to those employed in drawing the shelf, K, in plan.

270. *Execution.* As a line drawing, the plate explains itself. It would make a very beautiful shaded drawing and one that the careful student of the chapter on shading and shadows, would be able to execute with substantial accuracy, without further instruction.

THE END.

www.ingramcontent.com/pod-product-compliance
Lightning Source LLC
Chambersburg PA
CBHW021713210326
41599CB00013B/1639